Mohammed Alsaaq

Power System Harmonic Analysis Using ETAP

Mohammed Alsaaq

Power System Harmonic Analysis Using ETAP

LAP LAMBERT Academic Publishing

Impressum / Imprint

Bibliografische Information der Deutschen Nationalbibliothek: Die Deutsche Nationalbibliothek verzeichnet diese Publikation in der Deutschen Nationalbibliografie; detaillierte bibliografische Daten sind im Internet über http://dnb.d-nb.de abrufbar.
Alle in diesem Buch genannten Marken und Produktnamen unterliegen warenzeichen-, marken- oder patentrechtlichem Schutz bzw. sind Warenzeichen oder eingetragene Warenzeichen der jeweiligen Inhaber. Die Wiedergabe von Marken, Produktnamen, Gebrauchsnamen, Handelsnamen, Warenbezeichnungen u.s.w. in diesem Werk berechtigt auch ohne besondere Kennzeichnung nicht zu der Annahme, dass solche Namen im Sinne der Warenzeichen- und Markenschutzgesetzgebung als frei zu betrachten wären und daher von jedermann benutzt werden dürften.

Bibliographic information published by the Deutsche Nationalbibliothek: The Deutsche Nationalbibliothek lists this publication in the Deutsche Nationalbibliografie; detailed bibliographic data are available in the Internet at http://dnb.d-nb.de.
Any brand names and product names mentioned in this book are subject to trademark, brand or patent protection and are trademarks or registered trademarks of their respective holders. The use of brand names, product names, common names, trade names, product descriptions etc. even without a particular marking in this works is in no way to be construed to mean that such names may be regarded as unrestricted in respect of trademark and brand protection legislation and could thus be used by anyone.

Coverbild / Cover image: www.ingimage.com

Verlag / Publisher:
LAP LAMBERT Academic Publishing
ist ein Imprint der / is a trademark of
OmniScriptum GmbH & Co. KG
Heinrich-Böcking-Str. 6-8, 66121 Saarbrücken, Deutschland / Germany
Email: info@lap-publishing.com

Herstellung: siehe letzte Seite /
Printed at: see last page
ISBN: 978-3-659-58431-2

Zugl. / Approved by: Uxbridge, Brunel University, Diss., 2013

Copyright © 2014 OmniScriptum GmbH & Co. KG
Alle Rechte vorbehalten. / All rights reserved. Saarbrücken 2014

Acknowledgement

I would like to show a deep sense of gratitude to Allah (God) for providing me the blessings to complete this dissertation.

May I also take this opportunity to express my profound gratitude and deep respect to my project supervisor Dr. Mohamed Darwish, for his support and exemplary guidance all through my dissertation writing.

I would also like to thank Abraham Olatoke, for his valuable information and guidance, which helped me in completing this task through various stages.

Last but not least, I thank the, my mum, brothers, sisters and friends in Saudi Arabia for their continuous inspiration, without which this dissertation would not have been possible.

Abstract

Power system study and analyses are significant parts of power system engineering. Delivering good quality power is one of the major purposes of any electrical service company. In recent times, power quality has been affected in the power network caused by increasing the use in amount of DC devices, non-linear loads. Power system harmonics is one of the main causes for incorrect power quality. This has become a major issue and needs to be solved for power quality problems and harmonic analysis by analysing the electric power network to minimise the harmonics distortion current or voltage.

Power system harmonic analysis has been completed using ETAP software, and the entire process is explained in this project. ETAP software performs mathematical calculations of huge integrated power systems with magnificent speed as well, producing output reports. As a harmonic source a general load was demonstrated to inject harmonic current from side to side within the power network. Then, Harmonic Load Flow analysis was implemented and harmonic distortion identified to investigate the outcome of harmonic current. Lastly, an appropriate filter is advised for ETAP software to further improve the quality of the system.

Contents

Acknowledgement 3
Abstract 4
Chapter 1 10
Introduction 10
    1.1 Background 10
    1.2 Aim and Objectives 12
    1.3 Harmonic Analysis Features in ETAP 12
    1.3.1 Harmonic Load Flow 12
    1.3.2 Harmonic Frequency Scan 13
    1.4 Summary 13
Chapter 2 14
Literature Review 14
    2.1 Main Survey 14
    2.1.1 Linear and non-linear loads 14
        2.1.2 Types of non-linear loads 16
        2.1.3 Harmonics Current Flow 17
    2.2 Filters 18
        2.2.1 Active Filters 19
        2.2.2 Passive Filters 20
    2.3 When Harmonic Studies are performed? 27
        2.3.1 Harmonic study procedure 27
        2.4 Harmonic Analysis Module 28
        1.2 Harmonic Analysis 29
    2.5 Study Case Editor 31
        Info Page 31
        Plot Page 33
        Model Page 33
        Adjustment Page 34
        Alert Page 34
        2.5.1 Display Options 35
        Harmonic Analysis Display Option 35
        Harmonic Order Slider 36

Harmonic Frequency Slider	36
Report Manager	37
2.6 Summary	38
Chapter 3	**39**
Why Using ETAP	**39**
3.1 EATP Users	39
3.2 Harmonics analysis in ETAP	40
3.3 Optimal Capacitor Placement (OCP) in ETAP	41
3.4 Summary	42
Chapter 4	**43**
Harmonic analysis calculation methods in ETAP	**43**
4.1 Harmonic Analysis components modelling	43
4.2 Harmonic Indices	45
4.3 Harmonic Load Flow Study	47
4.4 Harmonic Frequency Scan	48
4.5 Harmonic Filter	48
4.6 Transformer Phase Shift	49
4.7 Standard Compliance	50
4.8 Summary	51
Chapter 5	**52**
Electric Network Simulation in ETAP, and Discussion	**52**
5.1 Power Network	52
5.2 Modelling the network components	53
5.2.1 Power Grid	53
5.2.2 Transformer	53
5.2.3 The Harmonic Source	53
5.3 Results of Balanced Load Flow Analysis	54
5.4 Results of Harmonic Load Flow Analysis	55
5.5 Results of Harmonic Scan	59
5.6 Harmonic Elimination	61
5.6.1 Harmonic Elimination Using Capacitor Bank (CB)	61
5.6.2 Harmonic Elimination using C Filter	65
5.6.3 Harmonic Elimination using Single Tune Filter + High Pass filter	67

5.7 Summary 71

Chapter 6 73

Conclusion and Limitation 73

    6.1 Conclusion 73

    6.2 Limitation 74

REFERANCE: 75

Table of figures

Figure 1 Voltage and current waveforms for linear 14
Figure 2 Voltage and current waveforms for non-linear loads.................. 15
Figure 3 Waveform with symmetrical harmonic components 15
Figure 4 Distorted current result in voltage distortion 17
Figure 5 Application of active filter at load................................ 19
Figure 6 Common passive filter configurations 20
Figure 7 Typical notch filter configuration 22
Figure 8 Series passive filter 22
Figure 9 Low-pass filter circuit (first order) 23
Figure 10 Frequency response of low pass filter 24
Figure 11 High Pass Filter 24
Figure 12 Frequency Response of High Pass Filter 25
Figure 13 Band Pass Filter... 25
Figure 14 Frequency Response of Band Pass Filter 26
Figure 15 C filter configuration 26
Figure 16 Main window (at edit mode) 28
Figure 17 Harmonic Analysis Toolbar 29
Figure 18 Harmonic Study Case editor 31
Figure 19 Create new Study Case 31
Figure 20 Harmonic Analysis Study Case (info page)........................... 32
Figure 21 Harmonic Analysis Study Case (plot page)Model Page................. 33
Figure 22 Harmonic Analysis Study Case (Model page) Adjustment Page 34
Figure 23 Harmonic Analysis Study Case (Alert page)2.5.1 Display Options 35
Figure 24 Display option (result page)....................................... 35
Figure 25 Harmonic Order Slider 36
Figure 26 Harmonic Frequency Slider.. 36
Figure 27 Study Case Editor Toolbar (Report manger) 37
Figure 28 Harmonic Report Manager Window..................................... 37
Figure 29 Harmonic page of the load and Library 45
Figure 30 ETAP Harmonic Filters 49
Figure 31 Transformer Editor, Tap page 50
Figure 32 4-bus power network 52
Figure 33 Load 2, Harmonic Page 54
Figure 34 Balanced Load Flow Results... 55
Figure 35 Load Flow Analysis Alert View 55
Figure 36 Harmonic Analysis Results 56
Figure 37 Harmonic Analysis Alert View....................................... 57
Figure 38 Harmonic Analysis Plot all buses (the voltage spuctrum)............ 57
Figure 39 Harmonic Analysis Plot all buses (the voltage wafeform) 58
Figure 40 Bus 3 Spectrum before the elimination 58
Figure 41 Harmonic Plot Bus 3.. 58
Figure 42 Result of Frequency Scan 59

Figure 43 Harmonic Frequency Scan Alert View... 60
Figure 44 Harmonic frequency scan, impedance angles (all buses) 60
Figure 45 Harmonic frequency scan, impedance magnitude (All buses)... 60
Figure 46 Harmonic frequency scan, impedance magnitude (All buses)... 61
Figure 47 Harmonic frequency scan, impedance magnitude (Bus 3) 61
Figure 48 Load flow analysis on left and Transformer Edit page for Tran 2 on right......................... 63
Figure 49 Harmonic Analysis after connecting capacitor bank.. 64
Figure 50 Voltage waveform before and after CB at Bus 3 64
Figure 51 Spectrums before and after the CB at Bus 3... 64
Figure 52 Frequency scan, Z angle before and after CB at Bus 3... 65
Figure 53 Frequency scan, Z magnitude before and after CB at Bus 3 65
Figure 54 C Filter Parameter page 66
Figure 55 The Network after connecting C Filter... 66
Figure 56 Voltage waveform before and after C filter.. 67
Figure 57 Spectrum before and after C filter... 67
Figure 58 Frequency scan, Z angle... 67
Figure 59 Frequency scan, Z magnitude 67
Figure 60 Harmonic Parameter and Filter Sizing Pages 69
Figure 61 Harmonic Analysis after connecting Single Tune Filter.. 70
Figure 62 Voltage waveform before and after notch filter at Bus 3... 70
Figure 63 Spectrum before and after notch filter at Bus 3 71
Figure 64 Frequency scan, Z angle before and after notch filter at Bus 3 71
Figure 65 Frequency scan, Z magnitude before and after notch filter at Bus 3 71

Table of figures

Table 1 The functions of active and passive filters 23
Table 2 Harmonics order 56
Table 3 Reduction in THD in each solution 72

Chapter 1

Introduction

1.1 Background

The main aim of any electricity company is to deliver electrical power in a pure sinusoidal form in a constant amount with better quality. However, this goal is sometimes hard to realise due to the harmonics current in the system that derive from non-linear load. These currents cause voltage and current distortion and badly affect the system's routine in different methods.

Harmonics happen as a result of modern electronics, and it has become progressively necessary to see their effect while making any additions or changes to an installation. Any electronics device using solid state variable speed motor drives, rectifiers, large numbers of personal computers (single phase loads), uninterruptible power supplies (UPSs) variable frequency drives (AC and DC) cause harmonics and make the voltages and currents distorted. Harmonics are created by non-linear loads that draw a current in the form of short pulses resulting in distorted sinusoidal. [13, 14]

In recent times, there has been an increase in the total amount of harmonics in the mains supply, as a result of increases in the quantity of power generation from renewable energy sources and more powerful electronics devices being added into the grid. Excluding harmonics currents caused by converters, which are used in renewable energy systems, is difficult. These converters are being used in renewable systems in order to reduce cost and increase reliability. But converters have a high level of total harmonics distortion (THD). Wind turbines have powerful electronic devices that create harmonics in the system, decreasing the quality of the system. As a result, the majority of renewable energy comes from wind turbines in western countries. [2, 11, 16]

Harmonics are the by-products of modern electronics. Harmonics are Sinusoidal voltages or currents having frequencies that are an integer multiple of the system's fundamental frequency. This project will be operated frequency (usually 50 or 60 Hz). [12]

Sources of harmonics can be classified as: [18]

- Traditional harmonic sources
  - Transformers.
  - Motors and generators.
  - Arcing devices.
- Modern power electronic harmonic sources
  - Electronic controls and switch-mode power supplies and office electronic equipment.
  - Thyristor-controlled devices.
    - Rectifiers.
    - Inverters.
    - Static VAR compensators.
    - Cycle converters.
  - HVDC power transmission systems.
  - Surge arresters.
  - Photovoltaic systems.
  - Electronic ballasts.
  - Welding machines.
  - Electrical Communication systems.

In this study, harmonics have affected the devices of the power system, the power quality, and the power factor. As a result, to undertake harmonics analysis relating to the power network, many power systems software packages can be used. This study will attempt to face the harmonics using ETAP software to model various power networks and perform harmonic analysis. ETAP has the facility to simulate harmonics' current and voltage course, identify harmonic problems, design and test harmonic filter and report harmonic voltage and current distortion limit. [12, 19]

1.2 Aim and Objectives

Analysing harmonics using ETAP software in the power system network and seeing how they can be eliminated are the main objectives.

To achieve this aim, these are the objectives:

- Learn how to use ETAP software.
- Provide an overview of harmonic analysis in ETAP software.
- Understand the difference between linear and non-linear loads.
- Recover the power system network, including power quality.
- Modelling harmonics courses in the power network.
- Design a suitable filter to minimise harmonics.
- Perform potential violations of distortion limits and frequencies in the power system network.

1.3 Harmonic Analysis Features in ETAP

Harmonics load flow and frequency scan are performed using detailed harmonics source models and power system component frequency models.

1.3.1 Harmonic Load Flow

All the information about load flow and harmonics can be produced within a Harmonics Load Flow report. The components of harmonics for voltage and current will be found in different ways such as harmonics load flow calculation and Total Harmonics Distortion (THD) and Individual Harmonic Distortion (IHD) will be compared with the limit identified through the operator. At fundamental frequency, the Harmonics Load Flow study performs a calculation of the load flow. The outcome of the fundamental load flow sets the base for the fundamental branch currents and bus voltage that finally are used to determine different harmonics indices. After that, every harmonics frequency and harmonics cause can exist in the system, and a good way to find direct load flow solution is a current injection method. [20]

1.3.2 Harmonic Frequency Scan

The best tool in this feature to examine the system resonance problem is the power station frequency scan program. A resonance condition might happen at several buses at some frequencies, and for that reason the system has inductive and capacitance components. Over voltage and over current will be observed when the resonance happens at a bus where a harmonic current is injected into the system. In addition, after evaluation and plotting the magnitude and the phase angle of the bus, the parallel resonance will be known. [20]

1.4 Summary

It is understandable that the major issue in power quality is the harmonic and powerful electronics devices that have affected system quality. The effect of harmonic current was analysed and a mitigation technique was applied to eliminate harmonic distortion. A passive filter and an active filter should be presented and selected based on the distortion in the system and the location because each filter has its features. The following chapter will take an overview of the difference between linear and non-linear loads, types of filters and Harmonic analysis module.

Chapter 2

Literature Review

2.1 Main Survey

Delivering sinusoidal voltage is one of the purposes of the electric value at an equally continuous magnitude all the way through their structure. As a consequence of the loads that generate harmonic currents on the system, it is difficult to achieve this. As the amount of harmonic generating loads has increased over the centuries, it has become progressively more obligatory to address the influence of harmonics before creating any additions or changes to an installation. In order to completely understand their impact, we need to take into consideration two significant thoughts of power system harmonics. Firstly, the nature of harmonic-current producing loads (non-linear loads). Secondly, the technique of the harmonic currents flow and the development of the harmonic voltages results. [1, 20]

2.1.1 Linear and non-linear loads

A linear load in an electrical power system is a component and can be known as the current is proportional to the voltage. By way of explanation, as we can see in the figure below, the current waveform has the same shape as the voltage wave (Figure 1). Linear loads include motors, heaters and incandescent lamps. [1, 20, 22]

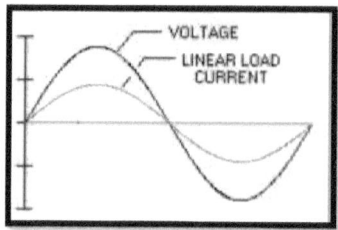

Figure 1 Voltage and current waveforms for linear (www.powerstudies.com/articles/Harm_Intro.pdf)

In the case of a non-linear load, the current waveform is not the same shape as the voltage wave (See Figure 2). Examples of non-linear loads include rectifiers (power supplies, UPS units, discharge lighting), ferromagnetic devices, adjustable speed motor drives, DC motor drives and arcing equipment. [1, 3, 20, 21]

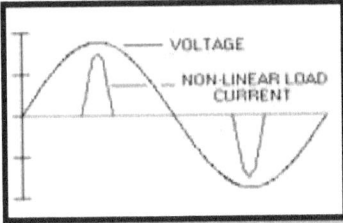

Figure 2 Voltage and current waveforms for non-linear loads (www.powerstudies.com/articles/Harm_Intro.pdf)

The current drawn by non-linear loads is not sinusoidal but it is periodic, meaning that the current wave looks the same from cycle to cycle. Periodic waveforms can be described mathematically as a series of sinusoidal waveforms that have been added together (See Figure 3). The sinusoidal components are integer multiples of the fundamental where the fundamental, in the United States, is 60 Hz. The only way to measure a voltage or current that contains harmonics is to use a true-RMS reading meter. If an averaging meter is used, which is the most common type, the error can be significant. [1, 3, 20]

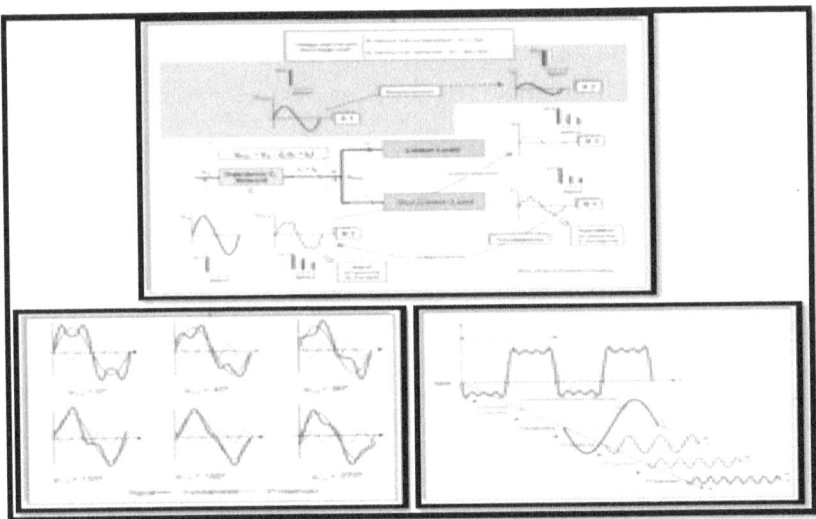

Figure 3 Waveform with symmetrical harmonic components (www.icrepq.com/ICREPQ'09/P1.pdf)

When waveform is symmetric it can be explained as when both positive and negative wave portions are identical. However, when the positive portions are different from the negative ones, it can be described as an unsymmetrical wave containing DC components. A large number of power system elements have been proven to be symmetrical. Components failures are considered to be a rare occurrence – in arc furnaces, for example, the device may possibly produce harmonics [1, 2].

2.1.2 Types of non-linear loads

This section will review non-linear loads types which cause harmonics in power systems. Three types of non-linear loads can be identified, these types include the following: [23]

1) Current source loads
2) Voltage source loads
3) A combination of current and voltage source loads.

- Current source

As a result of current waveform distortion on the AC side, the majority of non-linear loads in electrical systems are reported to operate as current sources. One of the most commonly-used examples is often found in the phase controlled thyristor rectifier, with filter inductance on the DC side of rectifier causing a DC current. The occurrence of current distortion is an issue when using an inverter to source AC power originating from DC. In another similar form, current distortion commonly forms as a result of high inductive provided by the rectifier to transform from AC power to DC. [23]

The harmonic generated solutions by current source loads are installed on the AC side of the converter serving non-linear loads such as passive shunt, active shunt, or hybrid filters.

- Voltage source

Examples of this can be observed when looking at capacitive filters that are installed in objects such as computers and household appliances as a voltage source; this is because these appliances contain diode rectifiers with capacitive filter at the DC link. A considerable degree of total harmonic distortion (THD) and poor power factor result, due to the fact that the loads draw discontinuous and non-sinusoidal currents in the process. Referring to these loads, instead of generating a current source, they generate a harmonic voltage source, due to the capacitive being installed in parallel. The voltage rectifier is far less dependent on AC system impedance, and this can be explained by the fact that the Diode rectifier behaves as a voltage source, since its harmonic amplitude is affected by the impedances of an AC side and source voltage [23].

Passive series, active series or hybrid filters can be referred to as being solutions used to abolish the harmonics caused by the voltage source that is being used.

- Combination of Current and Voltage source

Adjustable speed drives have been observed to act as a number of types of non-linear loads and also as some types of AC motor which can be current or voltage source, but this applies only to some devices.

The hybrid topology solution involves using active series with passive shunt filter to remove the harmonics [23].

2.1.3 Harmonics Current Flow

In this particular situation, the current will pass through the impedance which is also seen to connect between the load and power source, as seen in the circuit. This occurs when a non-linear load draws a current (Figure 4).

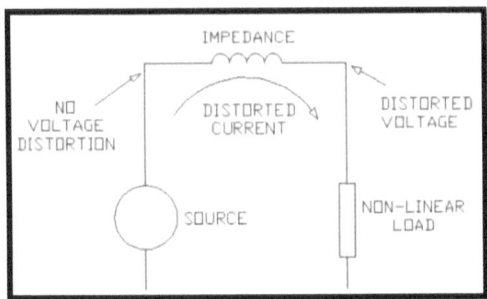

Figure 4 Distorted current result in voltage distortion (www.powerstudies.com/articles/Harm_Intro.pdf)

If the normal voltage is increased by adding the voltages sum, the result will be that the circuit will produce voltage distortion. By measuring both the source impedance and the harmonic voltage produced, the magnitude of the voltage distortion can be observed and identified. However, if the source impedance is not high, this will be reflected in the voltage distortion also not being high. In the scenario that a large portion of the load result and turns into non-linear (harmonic current increase), and/or when resonance condition overcomes (system impedance increase), these are the only instances when the voltage distortion can increase dramatically. As a final result, this will end in current flow, which also results in the production of harmonic voltage by impedance for each harmonic that is in the system. [1, 6]

In summary it can be said that harmonic currents pass through electrical power networks and also through distribution networks, resulting in voltage distortion being produced, because of the impedance related to the electrical network. Unwanted power is produced from harmonic currents and this takes up the system's capacity. Harmonic filters are introduced so that appliances can be saved and protected from this happening. [1, 17]

2.2 Filters

In order to terminate the harmonic distortion resulting from non-linear loads, filters are used effectively in the power system network. Harmonics filters are made up of a range of capacitors, inductors and resistors. Deflecting harmonic current to the ground is the purpose of the idea behind using the filter. It is also important to also realise that there can be harmonic filters contained within the system, each one with the purpose of deflecting the harmonics at a very specific frequency. Filters can be classified into two different types; these are labelled as active and passive filters. The active filter contains an amplifying device with the purpose of increasing the strength of the signal. Passive filters do not contain any amplifying devices, and their purpose is to terminate very specific harmonics, as well as being designed for specific loads. Active filters, however, are not limited in this sense and also have the ability to function with all loads. [24]

The benefits that harmonic filters bring can be listed as the following: [24]

- Reduce transformer loading.
- Protect electrical system.
- Minimise impact on distribution transformers.
- Increase system capacity.
- Decrease system losses.
- Improve power factor on non-linear loads.
- Reduce THD.
- Reduce neutral current, and protect neutral conductor.

2.2.1 Active Filters

This sort of filter is being used in low power network and this particular type known as active filter are used with regards for harmonics reduction in recent times. Not only are they based on developed power electronics, they are also a lot more expensive in comparison to passive filters, which are relatively cheaper. This filter can work self-sufficient of the system's impedance characteristics and also proves to not resonate with the system, and this can be regarded as one of the main advantages of this filter. Passive filters fail because of resonance problems, however, active filters are able to operate in multiple conditions. They are largely found in large distorted loads in power systems. Another advantage to this filter is that it is also capable of being used for resolving power quality issues like flicker, and can be employed to decrease many harmonics at the same time. [25]

One of the highest regarded values of the active filter is to substitute the ration of the waveform that is not found when looking at the current load in non-linear loads, as demonstrated in Figure 5. The voltage and also the current line are controlled by an electronic control, it is also used in power electronic devices in order to be able to track the load current and voltage and force them to be sinusoidal. As can be observed in the circuit displayed below in Figure 5, a capacitor or inductor can either be used with the filter and inductors that are being used to stock current to be added into the system within a precise time. The capacitor is also able to do the same function as the inductor. In addition, the current waveform can be seen by the system as more sinusoidal. This is true when the load current is not equal to the amount required by the non-linear load. [25]

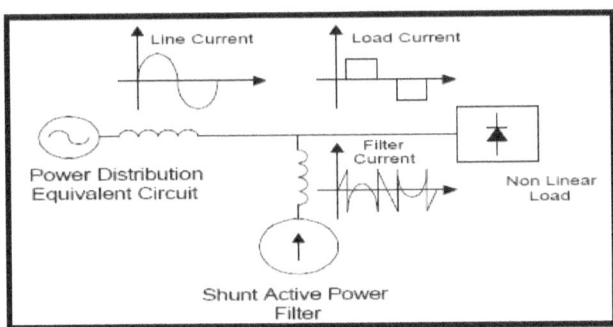

Figure 5 Application of active filter at load (http://web.ing.puc.cl/~power/paperspdf/dixon/37a.pdf)

Series active filter

Injecting a high impedance path to the current harmonic is a consequence of when a series active filter compensates for current harmonic distortion resulting from a non-linear load. By creating a voltage of the same frequency as the current harmonic that needs to be eliminated, this results are from high impedance presented by series active filter. [26]

Parallel active filter

Parallel active filter can compensate for current harmonics when inserting an equivalent but opposite harmonic compensation current. Consequently, parallel active filters will act similarly to the current source by introducing harmonic compensation created by load. When this filter is used in a suitable control scheme, it will be able to recompense the load of power factor. [26]

Hybrid active filter

When observed closely, the hybrid active filter can be seen as a combination of both active and passive filters; this is to eliminate the possibility of producing series or parallel resonance because it will increase the compensation characteristics of the passive filter. [26]

2.2.2 Passive Filters

Passive filters contain inductance, capacitance and resistance elements organized in such a way that they reduce harmonics. Passive filters are also widely found in the industry to decrease harmonic misrepresentation because of their clarity and cheapness. They are also used to force harmonic currents off the line or in order to stop the flow between the parts of the system; this action is achieved when altering elements to produce resonance at a selected frequency. As presented in Figure 6, certain kinds of common filter arrangements can be found currently being used in industries. Nevertheless, there is a downside of possibly interrelating in a harmful manner with the power system, and it is of very high importance to consider and observe all possible system interactions during the design stage. [9, 25]

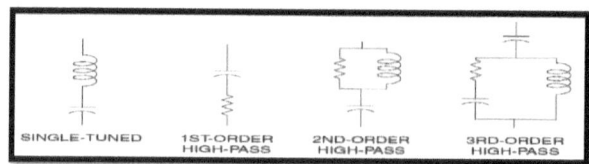

Figure 6 Common passive filter configurations [25]

Shunt passive filters

The most popular passive filter is the single tuned (notch) filter. It can be used for power factor alterations and correction and it is regarded as the highest in terms of cost effectiveness. It is also suitable for reducing harmonics. Harmonic currents change the flow path on the line through the filter, this results from when the notch filter is accompanied by an inductor and capacitor connected in a series to supply low impedance to a specific harmonic current and it is connected in shunt with the power system. In Figure 7, the typical configuration of a notch filter can be observed. [25]

It is vital to understand the side effects of the filter, which can create sharp parallel resonance at the frequency below that of the notch, and to take these side effects into consideration. It is essential that the resonant frequency occurs in a safe and secure area distant from all major harmonic and all other frequency components that can be formed by the load. In the event that the filter tunes precisely to the harmonic, the system parameter (capacitance or inductance) can be changed because of temperature or failure of results in shift with the parallel resonance higher into harmonic being filtered. Due to this, the filter must be tuned slightly less to that of the harmonic in order to be filtered. To be able to continue without failing because of the resonance problem, it is essential that the harmonic filters must be designed beginning with lower harmonic order. An example of this is when connecting the seventh harmonic in the network, it is common to need a fifth harmonic filter. The new parallel resonance created by the seventh harmonic alone is extremely close to the fifth harmonic which can then be labelled as being unsuccessful. [10, 25]

Finally, notch filters are also suitable for use in the correction of the power factor. It is essential that the notch filter must take into account the amount of the bus in to size the current carrying ability with regards to only the harmonic load produced, before it can be designed. It is possible to take the capacitor banks and change these into a filter by introducing inductance in series with the phases, as can be seen in Figure 7. [25]

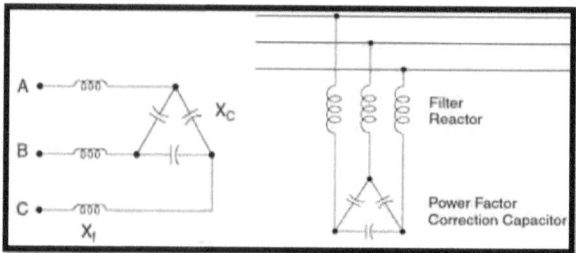

Figure 7 Typical notch filter configuration [25]

Series passive filters

Series passive filter connects in series with the load, unlike the notch filter, which is linked parallel with the power system. It is made up of capacitance and inductance, which are linked in parallel (Figure 8). The series passive filter will be tuned in order to deliver high impedance at selected harmonic frequency, doing so will block the current harmonic flow but only at the tuned frequency. It is essential that the filter be designed in order to deliver low impedance at a fundamental frequency, in doing so it will allow a fundamental current to flow with slightly extra impedance and loss. [9, 25]

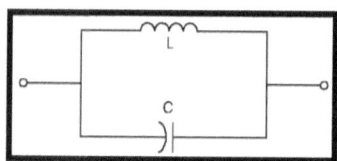

Figure 8 Series passive filter [25]

Series filters are employed particularly in single-phase circuits so as to block a single harmonic current; they are also used in a number of different types of circuit. It is essential that the series filter be designed in such a way that it can carry a full rated load current. In addition, it must have over-current protection. It is for this reason that the shunt passive filter is used over the series passive filter. All specific harmonic requirements separate series filter, however, this particular arrangement is able to generate large losses at the fundamental frequency, and this explains why series passive filters are not used for reducing multiple harmonics. [25]

Active Filters Functions	Series	Parallel
	• Eliminate voltage harmonics	• Eliminate current harmonics
	• Regulate the terminal voltage	• Reactive power compensation
	• Damp out harmonic propagation	• Balancing unbalance current
Passive Filters	All types	───
Functions	Compensate current and voltage harmonics	───

Table 1 The functions of active and passive filters [14]

In order to reduce harmonics, many different types of filters can be used. Different filters may tend to have different characteristics. A notch filter is therefore used to alleviate certain harmonic order, whereas a low pass filter or C filters are used in order to decrease widespread harmonics. ETAP is considered to be one of the most powerful software packages with the capability to design harmonic filters that reduce all harmonic distortion. [10, 25]

Low Pass broadband filter

A Low Pass Filter allows low frequency signals and rejects signals at frequencies above the filter's cut-off frequency. By connecting one resistor with one capacitor together in a series, a simple passive Low Pass Filter can be created, as seen in Figure 9. This kind of filter is well known as a first order filter as a consequence of it having a single reactive component in the circuit which is the capacitor. [22, 23]

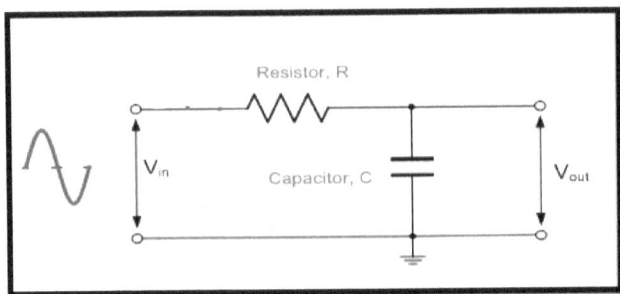

Figure 9 Low-pass filter circuit (first order) (http://www.electronics-tutorials.ws/filter/filter_2.html)

The capacitive reactance of a capacitor in an AC circuit is:

$$Xc = \frac{1}{2\pi fc} \text{ ohm's}$$

Where Xc is the capacitive reactance.

The circuit impedance Z formula:

$$Z = \sqrt{R^2 + Xc^2}$$

The output voltage of the filter is

$$V_{out} = V_{in} \times \frac{Xc}{\sqrt{R^2 + Xc^2}} = V_{in} \times \frac{Xc}{Z}$$

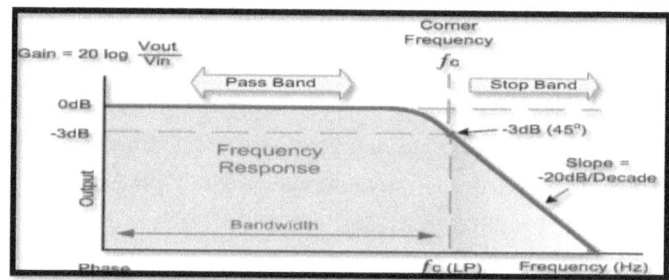

Figure 10 Frequency response of low pass filter (http://www.electronics-tutorials.ws/filter/filter_2.html)

High Pass broadband filter

A High Pass Filter circuit and Low Pass Filter circuit are different and opposite each other.[22]

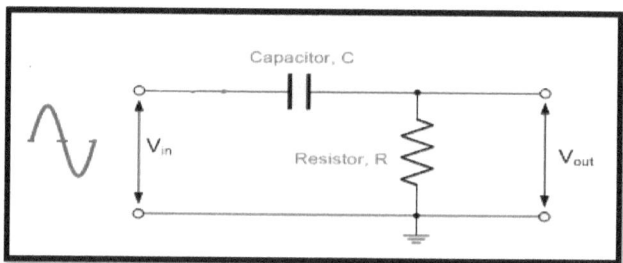

Figure 11 High Pass Filter (http://www.electronics-tutorials.ws/filter/filter_3.html)

Cut-off frequency, fc:

$$f_c = \frac{1}{2\pi RC}$$

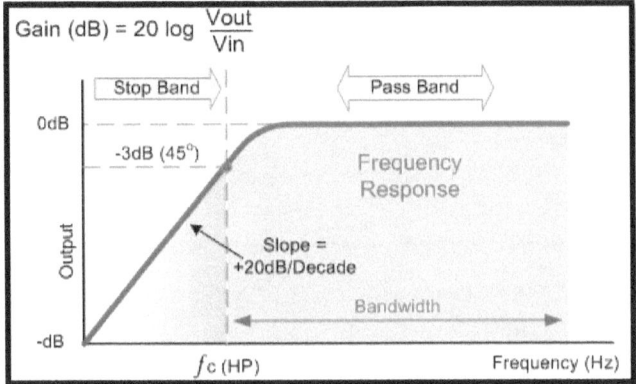

Figure 12 Frequency Response of High Pass Filter (http://www.electronics-tutorials.ws/filter/filter_3.html)

Band Pass filter

Another type of passive RC filter can be generated that passes a selected range by connecting together a single Low Pass Filter circuit with a High Pass Filter circuit, and can be either narrow or wide while reducing all those outside of this range. [22, 23]

A frequency selective filter known commonly as a Band Pass Filter is produced by this new kind of passive filter arrangement.

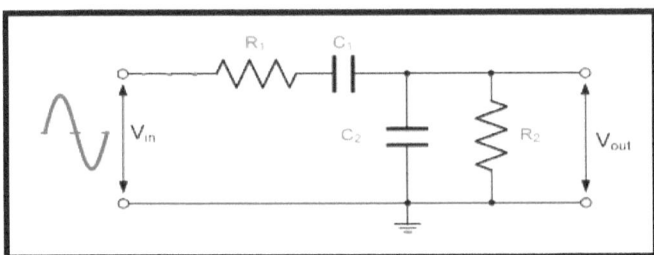

Figure 13 Band Pass Filter (http://www.electronics-tutorials.ws/filter/filter_4.html)

Band Width = $fH - fL$

Cut-off frequency, fc:

$$f_c = \frac{1}{2\pi RC}$$

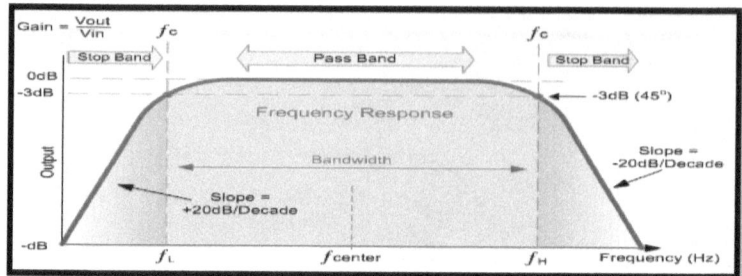

Figure 14 Frequency Response of Band Pass Filter (http://www.electronics-tutorials.ws/filter/filter_3.html)

C filter

Another type of low pass filter is a C filter. In industrial facilities and utility systems, the aim of this filter is the reduction of multiple harmonic frequencies. They are used to remove the harmonics produced by electronic devices such as cycloconverters and converters. [25]

The circuit of the C filter (Figure 15) is similar to a second-order high pass filter with an extra capacitor (C_a) in series with inductor (L_m). Also, this additional capacitor (C_a) is carefully sized regarding to its capacitive reactance cancels (L_m) at fundamental frequency, bypassing the resistor (R). Consequently, the losses are decreased at resistor (R), allowing the C filter to be adjusted at low frequency. [25]

Reducing high harmonic frequency by using the C filter is not sufficient. Therefore, at higher frequencies, a C filter and a notch filter are joined together to get more attenuation. The notch filter is associated with parallel resonance and this is one of the disadvantages of this configuration. This parallel resonance must be designated to save it from any harmonic current in the system. [25]

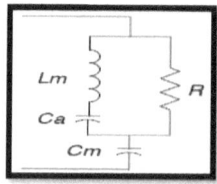

Figure 15 C filter configuration [12]

Limitation of passive filter

- Interaction with the power system.
- Changing characteristics (notch frequency) because of filter parameter variations.
- In particular cases it generates parallel resonance.
- It does not reflect the power quality of the system.
- The place of the filter and its parameters need to be change while waiting for finding an acceptable result in case of finding an inadmissible result. [23]

2.3 When Harmonic Studies are performed?

Harmonic studies are significant in term of recognising and explaining the size of harmonic problems. Therefore, harmonic studies are shown when: [25]

- Seeking solutions to an existing harmonic problem.
- Connecting great non-linear loads.
- Connecting big capacitor banks on utility distribution systems.
- Designing harmonic filter.
- Transforming power factor correction capacitor to filter.

2.3.1 Harmonic study procedure

The best process to begin the study is specified below: [25]

- Identifying the purposes of the study is one of the most important steps because this will provide the study with structure and direction. The objective in certain cases can be found and the existing problem solved, or it can be regulated if the new plant expansion comprises non-linear loads.
- It is a good idea to use a computer program such as ETAP if the system is complex, for example companies dealing with electrical grids, based on the best available information.
- Classify the harmonic source currents and the distortion of voltage on system bus in order to take measurements for the harmonic condition in the system.
- Adjust the computer model by the measurements achieved.
- Study the existing problem.

- Improved solution (filters) reflecting the interaction can be found, for instance resonance problems.
- Monitor the device to see the correct operations after installing the best appropriate solution for the problem,

As stated, this an ideal technique which is costly to apply due to the cost of engineers' time and tools. Also, only an expert can shortcut these stages from time to time to save money and time.

2.4 Harmonic Analysis Module

All the information and graphs in this chapter is taken from ETAP help [18]

ETAP 11.1.1 version is capable to analyse all the following:

1. Load Flow Analysis.
2. Unbalance Load Flow Analysis.
3. Short-Circuit Analysis.
4. Motor Acceleration Analysis.
5. Harmonic analysis.
6. Transient Stability Analysis.
7. Star-Protective Device Coordination.
8. Optimal Power Flow Analysis.
9. Reliability Assessment.
10. Optimal Capacitor Placement.
11. DC Load Flow Analysis.
12. DC Short Circuit Analysis.
13. Battery Discharging Analysis.

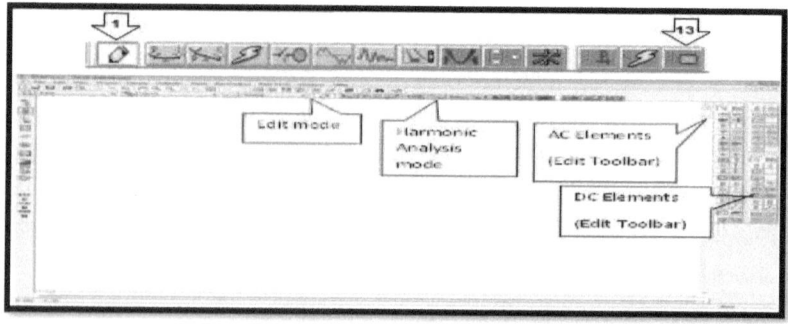

Figure 16 Main window (at edit mode)

1.2 Harmonic Analysis

After clicking the Harmonic Analysis from the toolbar mode the figure below will appear. This project measured the Harmonic Analysis in detail to overcome the harmonic distortion generated by a non-linear load. [18]

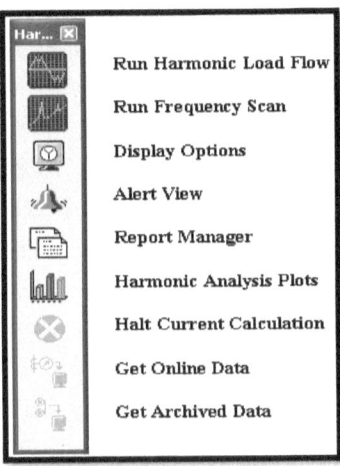

Figure 17 Harmonic Analysis Toolbar

- **Run Harmonic Load Flow**

    This option allows the performance of a Harmonic Load Flow study to specify the output report name for the output result. When calculation of Harmonic Load Flow is completed the output result can be viewed in the output report text and will appear in an online diagram. Another dialog box will also open where a Harmonic Order Slider will appear that can be calculated and this will show the result for different harmonic order in an online diagram. [18]

- **Run Frequency Scan**

    The Run Frequency Scan option will implement a harmonic frequency scan study and evaluate the magnitude and the impedance angles to plot the result as a spectrum format and waveform. These results can be observed in an output report text on the online diagram and formats plotted after the calculation is completed. The Slider of Frequency is obtainable to provide the result in every single harmonic frequency. [18]

- **Display Options**

Display Option allows the user to display under Harmonic Analysis Study mode. A dialog box will appear to customise the results to be displayed in the online diagram. [18]

- **Alert View**

  Alert View has all its equipment with critical and marginal violations based on the settings in the study case. When a Harmonic Load Flow and Frequency Scan is performed, or a Harmonic Frequency Scan, this option will allow the user to open the Alert View (IEEE519 set standard in ETAP). [18]

- **Report Manager**

  This is to allow the user to choose a format and display the best power output reports. These reports are provided in Crystal Report Viewer, PDF, MS Word, Rich Text, and MS Excel formats. Some of the predefined reports originate from here in Complete, Input, Results and Summary pages respectively. [18]

- **Harmonic Analysis Plot**

  This option allows users to plot for instance Buses, transformers. The plot file name is similar to the output text file shown in the Output Report pull down list. After clicking this button a dialog box will appear to choose the elements. [18]

- **Halt current calculation**

  If a Harmonic Load Flow or a Harmonic Frequency Scan is initiated, this button is usually disabled and shows a red stop sign and turns into enabled. Current calculation would be stopped by pressing this button. This option is normally used in a real time process. [18]

- **Get Online Data**

  From the online presentation, the user can copy the online data to a current presentation when ETAP is installed in a computer and has online features. [18]

- **Get Archived Data**

  Get Archived Data and Get Online Data act in the same way. When ETAP has online features, users can copy the archived data to a current presentation when ETAP is installed in a computer. [18]

2.5 Study Case Editor
2.5.1 Harmonic Analysis Study Case Editor

The Harmonic Analysis Study Case Editor includes loading conditions, solution control variables, and a variety of options for output reports. ETAP lets users make and save an infinite number of study cases. The calculations of Load Flow are conducted and stated along with the setting of the study Case chosen in the toolbar. Switching between the study cases without resetting can be done by the user easily. By clicking on the Study Case option from the Harmonic Analysis mode, the user can access to the Harmonic Analysis Study Case Editor. [18]

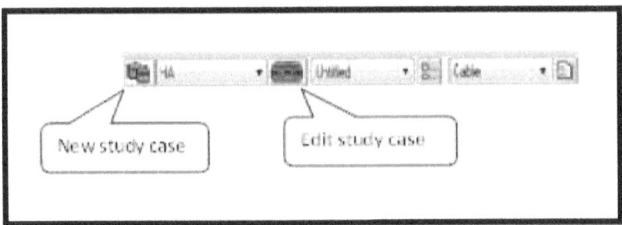

Figure 18 Harmonic Study Case editor

After clicking the new study case button, the user will also be able to create a new study case and after that it can be edited according to particular requirements. The button of Edit Study Case involves five tabs where the user can edit for efficient harmonic analysis. The five tabs under study case editor are Info, Plot, Model, Adjustment and Alert. [18]

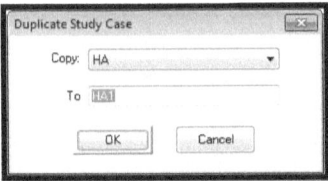

Figure 19 Create new Study Case

Info Page
ETAP provides this page to the user to choose general solution parameters, report options loading conditions, and study case information. [18]

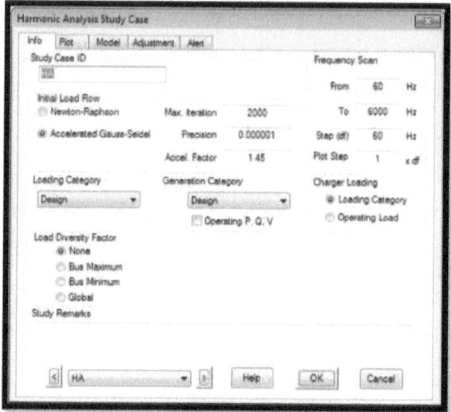

Figure 20 Harmonic Analysis Study Case (info page)

- **Study Case ID**

  This study case ID is obtained in this entry field and has ability to rename the Study Case by removing the previous ID one and entering a new one. [18]

- **Fundamental Load Flow**

  These options are applied to both Harmonic Load Flow and Harmonic Frequency Scan studies, and used for the fundamental load flow calculation solution control. [18]

    o Max. Iteration
    o Precision
    o Accel. Factor

- **Frequency Scan**

  These values are used for only harmonic, frequency scan calculations. [18]

- **Fundamental Loading**

  In this part, the system loading conditions can be specified for the fundamental load flow calculation. The harmonic load flow and the harmonic frequency scan calculations will be affected by the fundamental loading conditions. [18]

- o Loading Category
- o Generation Category
- **Charger Loading**

  This option to use the P and Q identified in the Loading Category section of the Charger Editor for chargers. [18]

  - o Load Category
  - o Operating Load
- **Load Diversity Factor**
  - o None
  - o Bus Maximum
  - o Bus Minimum
  - o Global

Plot Page

The components that will be displayed in the online diagram and plot format can be selected by the user. The Harmonics Load Flow and Harmonic Frequency Scan studies are applied by the components previously selected by the user and the components that are being used in the network will be listed in the Device Type. A list of devices is available for a given Device Type, but only after the user selects the specific component. The user now has the ability to select any components that need to be plotted. [18]

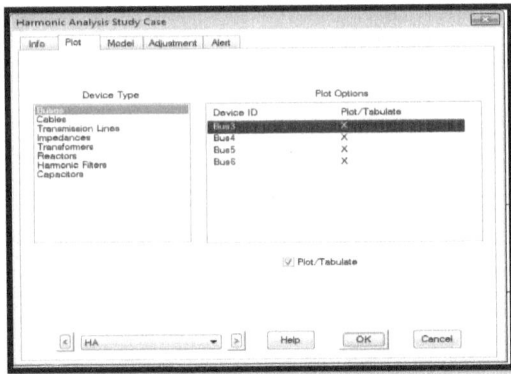

Figure 21 Harmonic Analysis Study Case (plot page)Model Page

The following page is provided for the user in order to select modelling methods for a variety of different component types. The Exclude Harmonic Source group specifies universally what type of components the operator does not need to model as harmonic sources. The Harmonic Load Flow and Frequency Scan are affected in terms of their results. The Transmission Line/Cable tap can be organized into two different options, short line and long line transmission models. [18]

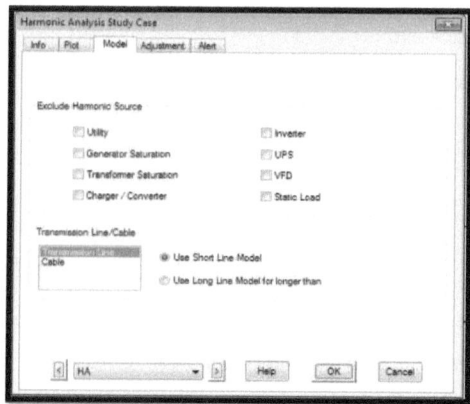

Figure 22 Harmonic Analysis Study Case (Model page) Adjustment Page

The Adjustment Page gives the user the ability to specify tolerance adjustments to terms of length, equipment resistance, and impedance. All tolerance adjustments can be applied centred upon individual equipment, percent tolerance setting and also on the globally specified percent value. [18]

Alert Page

When observing the Harmonic Analysis toolbar it can be seen that the Alert Page is also available. All alerts gathered from Harmonic Load Flow Analysis originate from the base of the setup of this page to inform the user of all abnormal conditions, making the feature of very high importance. There are two different types of simulation alerts, these are called Critical and Marginal and they are provided by the ETAP. The two of them are different when looking at percent value conditions to determine if an alert should be generated. [18]

Power System Harmonic Analysis Using ETAP

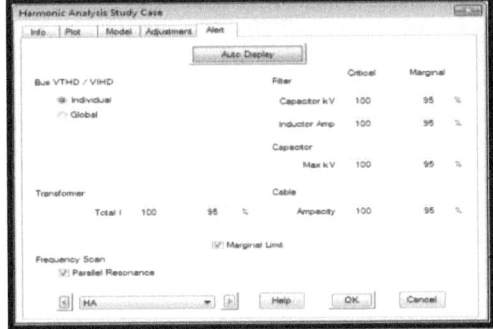

Figure 23 Harmonic Analysis Study Case (Alert page) 2.5.1 Display Options

Harmonic Analysis Display Option

The Harmonic Analysis Display Option is provided in the ETAP to the user to identify the data shown in the online diagram. It contains four pages, Results, AC, AC-DC and Colours. [18]

Figure 24 Display option (result page)

- For the Results page, after clicking on the check box it will show the voltage in the diagram as kV or percentage. The information on the Brunch Current and other

Mohammed Alsaaq (0926125)   Page 35

Brunches acquired from Harmonic Load Flow and Frequency Scan calculations will be contained within the Brunch Current check box. The Total Harmonic box is used to state the displayed information for Buses and Brunches in (RMS) or (ASUM) (Arithmetic Summation). The frequency Scan group is presenting the bus driving point impedance (Z impedance) or bus driving point impedance phase angle (Z angle). [18]

- The (AC), (AC-DC) page has choices for showing information annotations. In (AC) user can display the generator information and power grid. The (AC-DC) can show the components for instance inverter and charger. [18]

Harmonic Order Slider

After running the Harmonic Load Flow Analysis, the Harmonic Order Slider box will be seen. Total, 1 (fundamental frequency), and h (harmonic order from 2 to 73) are the three sections of the Slider. A mouse pointer is used to change and set the slider at different level. [18]

Figure 25 Harmonic Order Slider

Harmonic Frequency Slider

As the info page of the Harmonic Analysis Study Case Editor mentions, the values of the frequency scan are selected by the user. For the selected frequency, the Harmonic Frequency Scan Slider shows the phase angle and the magnitude of the bus driving point impedance. [18]

Figure 26 Harmonic Frequency Slider

Report Manager

There are three formats for the Harmonic Analysis Study reports: Crystal Report, online diagram displays and plots. In order to get the Report Manager, there are two ways, the first being the shortcut by clicking on Report Manger option, as in the figure below. [18]

Figure 27 Study Case Editor Toolbar (Report manger)

The other way is by choosing the Report Manager placed at the Harmonic Analysis toolbar. There are five different formats as shown in the figure below to view the reports; Viewer, PDF, MS Word, Rich Text Format, and MS Excel. [18]

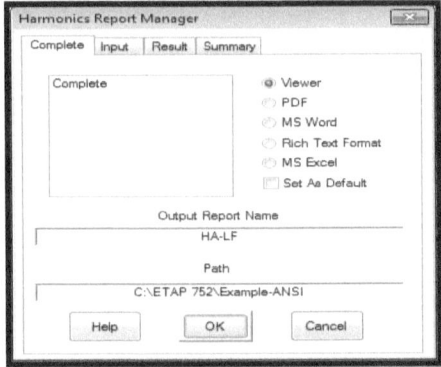

Figure 28 Harmonic Report Manager Window

The user can press on Report Manager Option to view the completed page for the results for the entire network. The Input report page provides the formats for many input data. The result page can show the formats for different calculation results. Finally the summary page provides two formats for many summaries, including input data and calculation results. [18]

2.6 Summary

Ultimately, this chapter covered three main topics, including the types of load and the difference between them. In addition, it discussed how, as a solution for harmonics, filters are introduced and how it is essential to understand how every single filter works. Finally, ETAP was presented in order to analyse the software's performance, specifically in harmonic analysis. The next chapter will analyse harmonics in networks, using the data obtained from this chapter, and the reasons for choosing ETAP for this project.

Chapter 3

Why Using ETAP

3.1 EATP Users

Over recent years electrical engineers have been concentrating on power system studies by using software tools, for example Electrical Transient Analyser Program (ETAP software). In more than 3,000 companies manufacturing around the world and electrical utilities power systems, The Electrical Transient Analysis Program (ETAP) is used by engineers. ETAP electrical power equipment and facilities are used in all periods of power development, beginning from power generation, to transmission and from distribution to operation. ETAP users are listed by industry containing; engineering consulting, data centres, oil and gas, generation plants, nuclear generation plants, transmission line, government and military facilities, metals and mining, manufacturing, renewable energy, transportation, and education. In addition, there are two reasons for choosing this kind of software, including the fact that Brunel University has ETAP software in some of their appliances, and because it is new software with precise calculation methods. As a result, ETAP offers an educational version of this software package for mentors, university electrical labs, and students in electrical engineering departments at colleges and universities. ETAP is provided at Brunel University to teach students how to design, model, and analyse electrical power systems. [5, 27]

To quote the ETAP official site in data centres and mission critical services "IT operations are a crucial aspect of most organizational operations. One of the main requirements is ultra-high reliability and availability of the electrical power system that must meet stringent 24/7 operating criteria to maintain continuous functionality and minimise costly unscheduled downtime. While security and availability come first, energy usage is becoming a focus when considering long-term operations." [27]

On the other hand, there is software similar to EATP in some features such as Power Factory software. It is common and is considered to be old software, used for more than 25 years, which is able to model generation, transmission, distribution and industrial grids. Moreover, this kind of software is reliable. Power Factory software and ETAP software have some similar features. Nevertheless some differences will become apparent once the Harmonic analysis is considered. [4, 27]

There are three main variances among Power Factory and ETAP. Firstly, Power Factory is able to run Harmonic Load only while ETAP has the skill to run Harmonic Load Flow and Harmonic Frequency Scan Flow analysis. The second difference is the calculating process of Harmonic Load Flow analysis. Power Factory utilises only the Newton-Raphson method, however ETAP uses two kinds of calculating methods, the Newton-Raphson and the Accelerated Gauss-Seidel method. Lastly, both software versions have different systems to generate harmonics in the system. In Power Factory a harmonic source can be selected from different techniques, whereas a harmonic source can be selected from the library in ETAP software. [4, 27]

Performing Plot in Power Factory is complicated, whereas in ETAP it involves using one tool. Also, it is very easy to design all the six passive filters types in ETAP by placing them in the filter size page after taking certain limits from the analysis. On the other hand, Power Factory uses different types of passive filters. Up to now the user needs to do some manual calculation in some situations. The library in ETAP has an extensive variety of tools, however in Power Factory some of these tools do not exist. [27]

Ultimately, ETAP is more common and popular than Power Factory. ETAP software has many good points including the fact that it can be used easily in term of Harmonic analysis and designing filters, plus it uses hardly any tools for extensive analysis and that enables operators to understand it more quickly compared with Power Factory. In the next section, harmonics analysis in ETAP will deliberate the two main features in ETAP concerning the Harmonic analysis to identify all its capabilities. [4, 27]

3.2 Harmonics analysis in ETAP

In this project ETAP is used due to its capability and flexibility in terms of classifying harmonic glitches, raising system efficiency, designing and checking filters and reporting any that go beyond distortion bounds. There are two ways to perform clearly the harmonic impact on the system, harmonic load flow and frequency scan calculations. It also computes the system impedance from magnitude and angle at any designated bus and displays the fundamental load flow result. The outcome of calculations can be found in detail as graphs for each harmonic order and moreover can be shown as spectrum and waveform plots. [15]

Significant features of harmonic analysis: [15]

- Analysis harmonic load flow
- Harmonic frequency scan
- Filter design and sizing
- Inter-harmonic filter modelling
- Calculate harmonic limits automatically

Capabilities of harmonic analysis: [15]

- Able to design single tuned, high pass and band pass filters (passive filters).
- Shifting resonance points by generating filter
- Achieve to 71st harmonic by modelling up
- Classify and analyse telephone interference difficulties
- Harmonic source library consisting of harmonic source to be displayed
- Recognise resonance conditions
- Temperature-dependent line and cable resistances.

3.3 Optimal Capacitor Placement (OCP) in ETAP

In Total Harmonic Distortion (THD), it is well known that capacitor bank can improve the voltage profile and power factor which result in reduction. OCP has an ability to perform this plus reduce the long term operation. The user can show the result as graphs and control the capacitor placement method. An exact calculation can find bank sizes and the best sites. In reactive power, it likewise can control the saving throughout the planning period outcome of loss reduction. [15, 29]

Significant features of OCP: [15]

- Optimal site and bank size
- Separate source
- Easy and flexible constraints
- Capacitor control technique
- Operation budgets and minimise installation
- Branch capacity issue and price saving
- Power factor impartial and voltage

- Minimum, maximum and average loading
- Realise capacitor impact on the system

Capabilities of OCP: [15]

- Discover the best location for the capacitor and bank size
- Find a global optimal result by using Genetic algorithm
- Using maximum load to calculate maximum capacitor size.
- different loading types
- Considering saving during whole planning period
- Determine capacitor magnitude
- Able to plot the loss reduction saving for the period of the planning period, the total cost of capacitor operation cost, and yearly saving.

3.4 Summary

The features of OCP and harmonic analysis make ETAP reliable and cost effective compared with other electrical power software. When dealing with harmonics, there are two main features which must be considered in ETAP software, Harmonic Analysis and OCP which run analysis about harmonics.

The following chapter will debate ETAP calculation methods to know two ways including software standards and how to model harmonic source.

Chapter 4

Harmonic analysis calculation methods in ETAP

The best tools for modelling many power system components and devices are provided by the ETAP Harmonic Analysis component to contain their frequency need, non-linear loads, and extra features under the presence of harmonic sources. For power system harmonic analysis, the Harmonic Load Flow method and the Harmonic Frequency Scan method are two ways of applying this module and these are the best-known and the most powerful methods. Also, with the industrial standard limitations between the methods above, different harmonic indices are calculated and the problems of potential power quality are compared, together with problems of security related to harmonics, which can be easily discovered. The reasons for these problems can be acknowledged and several developments, for example power factor correction and harmonic filter, can be tested and lastly confirmed. [2, 21]

4.1 Harmonic Analysis components modelling

For well-organised harmonic analysis, frequency characteristics and non-linear loads, the components of the power system are demonstrated in different ways depending on their behaviour and their nature. These components must be modelled in a number of ways. There are two ways to form the non-linear loads in a power system, either by injecting harmonic currents into the system or putting on harmonic voltages at known points. Therefore, if normal power sources, for instance generators or power grids, contain harmonic components in their fixed voltage, this means they would be modelled as a voltage source with harmonic frequencies. [6, 8]

Non-linear loads in ETAP that can be generated as harmonic current source are: [8]
- UPS
- Transformer
- VFD
- Static Load
- Charger/Converter
- SVC

The harmonic current will be injected into connected buses such as static load, converter and VFD if they are modelled as a harmonic current source. As a harmonic current source, the transformer can be modelled because it generates harmonic current when it is lightly loaded. At the primary side, the harmonic current source is positioned; however, as soon as a triple n^{th} harmonic current stated for a transformer and the winding and ground connection of the transformer do not let the nth harmonic current flow in the primary winding, it will be considered as the location of harmonic current source in the secondary side. [7, 8]

A UPS injects harmonic current into the connected bus. There are two major cases for the UPS if it is modelled as a load and if it is modelled as a branch, both AC input bus and AC output bus will inject current. For that reason, the path inside the UPS will be unlocked in harmonic load flow calculations from AC input bus to AC output bus. [8, 28]

The source will be specified in terms of the percentage of the components to cause the magnitudes of the harmonic current. At the Harmonic page of the component, Harmonic Page can select the appropriate harmonic current library and Harmonic Library Quick Pick Editor to model components as harmonic current source using ETAP. [8]

In ETAP the components that are modelled as harmonic voltage source are specified in the points below:
- Power Grid
- Synchronous Generator
- Static Load
- Charger/Converter
- Inverter

When the voltage distortion instead of current distortion is caused by the charger/converter, the inverter and static load, they can be modelled as a harmonic voltage source. In power sources the harmonics current sources in ETAP, whether power grid or synchronous generator, can be allocated as a voltage source of harmonics.

To model any of these devices above as a harmonic voltage source, the applicable source should be selected from the library through the Harmonic Page of the device. The subsequent (Figure 29) illustrates the Harmonic Page with source type (voltage/current) and model (six-pulse/twelve-pulse rectifier). The magnitudes of the harmonic voltage are in percentage of the voltage. [8, 28]

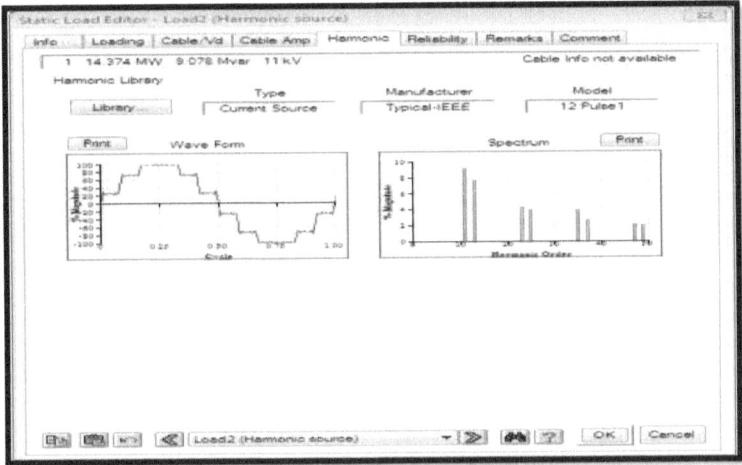

Figure 29 Harmonic page of the load and Library

4.2 Harmonic Indices

In Harmonic Indices, measurements are defined below, and the special effects of the harmonic are shown:

- **Total Harmonic Distortion (THD)**

    Total Harmonic Distortion (IHD) or Harmonic Distortion Factor (HDF) have the same meaning, the main aim of THD is to evaluate the level of harmonic distortion in both voltage and current. THD displays the ratio of the mean-square-root of all harmonics to the fundamental components, and the ideal system is equivalent to zero. [8]

    $$THD = \frac{\sqrt{\sum_{2}^{\infty} Fi^2}}{F_1}$$

    Where:

    Fi : The amplitude of the ith harmonic,

    F_1 : The fundamental components.

- **Individual Harmonic Distortion (IHD)**

    Individual Harmonic Distortion (IHD) computes the ratio of the harmonic component to the fundamental component. The value obtained thus is used to track every single harmonic individually and test its magnitude. In addition, it is possible to design single tuned filters. [8]

$$IHD = \frac{Fi}{F_1}$$

- **Root Mean Square (RMS)**

   In this part, RMS is the square root of the sum of the squares of the magnitudes of the fundamental along with all harmonics in the system. The fundamental component should be equal to the total RMS for the ideal situation once there is no harmonic. [8]

$$RMS = \sqrt{\sum_1^\infty Fi^2}$$

- **Arithmetic Summation (ASUM)**

   In this section, ASUM is used to calculate the summation of the fundamental magnitudes and all harmonics. It adds the magnitudes of all components directly in order to get the crest amount of the voltage and current. [8]

$$ASUM = \sum_1^\infty Fi$$

- **Telephone Influence Factor (TIF)**

   Telephone Influence Factor (TIF) is a difference of the THD with many weights assumed to every one of harmonics based on the quantity of interference to an audio signal in the same frequency range. In communication systems, the current of TIF has further significant impact. [8]

$$TIF = \frac{\sqrt{\sum_1^\infty (Wi * Fi)^2}}{\sqrt{\sum_1^\infty Fi^2}}$$

- **I*T Product (I*T)**

    I*T Product is defined as the product of current components and weighting factors. [8]

    $$I * T = \sqrt{\sum_{1}^{H} (I_h\, T_h)^2}$$

Where:

- h : Harmonic order (h=1 for fundamental),
- H: Maximum harmonics order to account,
- I_h : Current component,
- T_h : Weighting factor.

4.3 Harmonic Load Flow Study

Harmonic Load Flow has many features. Users can simulate harmonic current and voltage sources, reduce nuisance trips, categorise harmonic problems, design and test filters and report harmonic voltage and current distortion limit violations in ETAP's Harmonic Analysis module. At fundamental frequency, load flow calculation is carried out by the Harmonic Load Flow Study. For the fundamental bus voltages and branch currents, the consequence of the fundamental load flow sets a base to analyse different harmonic indices which are used later. After that, a load flow solution is produced for each harmonic frequency by using the current injection method at which any harmonic source exists. The characteristic harmonics are considered starting from the 17th to the 73rd and low order frequencies are measured starting from the 2nd to the 15th. Impedance of the components will be corrected based on the components types and the harmonic frequencies. [7, 8]

Branch current will be obtained by placing the components of harmonics in the voltage bus and, in keeping with the result achieved, all harmonic indices will be determined after running the harmonic load flow analysis. At bus editor, the limit values specified by the user are compared with the calculated bus THD and IHDs. If warning flags are seen in the text that appear in the next harmonic load flow report, this means the results have exceeded their limit. The Newton Rephson method or the Accelerated Gauss Seidel method are used to calculate this load flow. This report of harmonic load flow generates input data of the system, the results of fundamental load flow, system harmonic information, and the bus voltage and branch current with all harmonic contents. The Harmonic Load Flow Slider and the Harmonic Display Options Editor are used to show the results directly from the online diagram. A plot for bus voltage and brunch current can be demonstrated to see the voltage and current waveforms in time domain, and the harmonic spectrums in a bar chart. [8]

4.4 Harmonic Frequency Scan

The resonance condition in the power system is one of the most significant sources of concern with harmonics. As a consequence of the existence of both inductive and capacitive modules at certain frequencies in the power system, resonance conditions might happen at a number of buses. Once the resonance happens and if the harmonic current is injected, it will generate an overvoltage condition at the same bus. If a harmonics source exists at the frequency where the impedances are the same, parallel resonance causes problems. [8]

The ETAP Frequency Scan program is used to investigate resonance problems. It computes and plots phase angles and the magnitudes of the bus over frequency and it will be easy for the users' option to tune their filters' parameters and test the final result to find any parallel resonance condition and its resonance frequency. [8]

The frequency scan reported results of harmonic load flow comprises input data of the system, the results of fundamental load flow results, and information about resonance up to the point selected by the user. The resonance location can be plotted as waveform based on the results obtained. [8]

4.5 Harmonic Filter

Harmonic filters are commonly used to reduce the harmonic problems at the lowest level. A properly designed harmonic has the ability to stop harmonic currents from being injected into the system, and at tuned frequency, some filters deliver low impedance paths that eliminate parallel resonance, such as notch filter. [8]

Nowadays, ETAP has provided the most common filters used in the power industry and a Filter Sizing program can be found at Harmonic Filter editor for the Single Tuned filter type, which helps the user to consider the loading of harmonic filters and to optimise the filter parameters, depending on the operation criteria. Capacitor Max. kV and inductor Max. I are the two important loading values, where these values are available in Harmonic Filter editor. The voltage drop across capacitor is used to compute the capacitor Max. kV which is a peak value, bearing in mind the current flow through the inductor to calculate the inductor Max. I which is a root mean square (RMS) value. The fundamental and harmonic values for the voltage and current are included to get Max. kV and Max. I values. There is a check box in the Harmonic Analysis Study Case Editor if the Filter is Overloading. Between the calculated values and the identified values comparisons will be completed by using Harmonic Load Flow Study if the information page is checked in the Filter Overloading report. The percentage of overloading will then be calculated and reported. [8]

In ETAP, here are types of filter respectively as in the (Figure 30) from left to right:

- Single Tuned Filter
- By-Pass Filter
- High Pass Filter (Damped)
- High Pass Filter (Undamped)
- 3rd Order Damped Filter
- 3rd Order C Type Filter

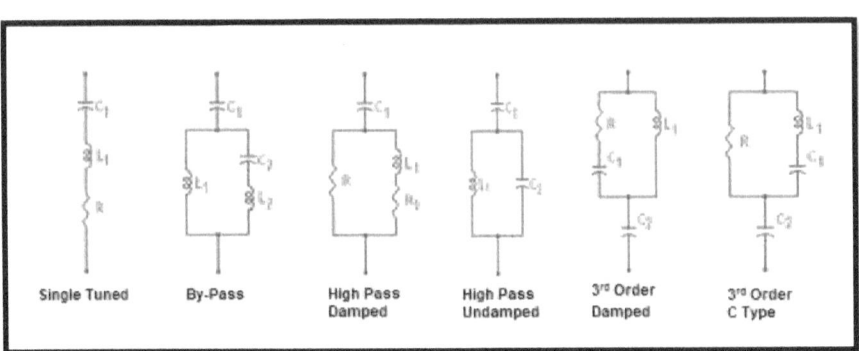

Figure 30 ETAP Harmonic Filters

4.6 Transformer Phase Shift

Power system harmonics can be minimised by proper tap setting. Transformer phase-shift is one of the uses of ETAP Harmonic Analysis modules and it is valuable to cancel some harmonics results in order to get a better power quality system. The ETAP uses transformer phase-shift in the harmonic load flow to regulate the network impedance phase angle. In the Transformer Editor Tap page, transformer phase shift is quantified based on transformer connection convention. As shown below, the phase-shift is known as Standard Positive Sequence connection and Standard Negative Sequence connection. On user defined part, the user has the ability to identify special-phase shift for a transformer by entering the degree of phase-shift to reduce certain harmonics. [8]

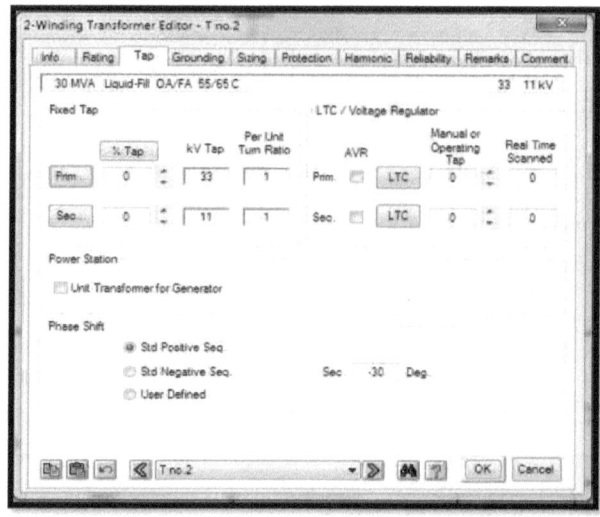

Figure 31 Transformer Editor, Tap page

4.7 Standard Compliance

Harmonic Analysis module in the ETAP software acts in accordance with the latest version of the following standards:

- IEEE standards 519-1992, IEEE Recommended Practices and Requirements for Harmonic Control in Electrical Power Systems.
- ANSI/IEEE Standard 399-1997, IEEE Recommended Practice for Industrial and Commercial Power System Analysis.
- IEEE Standard 141-1993, IEEE recommended Practice for Electric Power Distribution for Industrial Plants. [8]

4.8 Summary

This chapter has highlighted the method of harmonics analysis using ETAP software and that support ETAP. The harmonic load flow, harmonic frequency scan computing harmonic indices at given buses and branches, components modelling and finding the problems with existing harmonics have been discussed in this section. Chapter number five will analyse harmonics in a network based on data and facts obtained from this chapter and will then use different methods to decrease the THD.

Chapter 5

Electric Network Simulation in ETAP, and Discussion

The following section contains information that has been obtained from theories as well as the principles of ETAP, and will be used in electrical power networks in order to eliminate the harmonics in the system, meaning that quality will improve significantly in relation to the system. Distortions are created in the network, resulting from harmonic components which are modelled in order to create harmonic current and action must be taken in order to be unaffected by this problem. Possible solutions could include proper filter design, phase shift, or capacitor bank. The cost and the reliability of the solution chosen should be considered in order to find an optimum solution.

5.1 Power Network

The network represented in (Figure 32) consists of an online diagram with the objective of performing a harmonic and load flow analysis. The harmonic source can be found at Bus 3, and the network consists of 4-buses. In order to find exactly where the impact of the harmonic on buses is and the surpassed amount of THD, an investigation will be performed using the harmonic analysis tools.

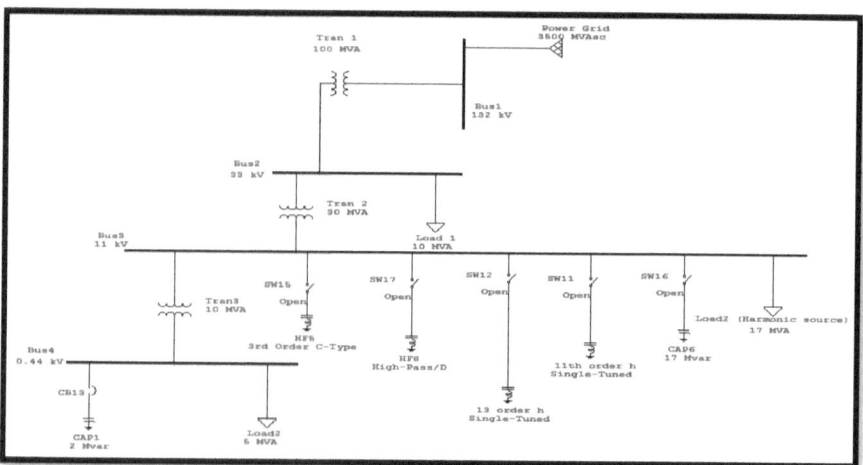

Figure 32 4-bus power network

5.2 Modelling the network components

5.2.1 Power Grid

With regards to the power grid, it is essential that the rated (kV) is stated by the user which in this network 132 kV, this particular value is also found being used as power grid base. However, this value of short-circuit (MVA) regarding the three phase is found by (MVA_{SC} = $\sqrt{3}$ ×kV×I. X) this is the ratio for the positive and zero sequence impedance.

5.2.2 Transformer

Both the primary and secondary values of the 2-winding transformer are stated by the user in (kV). A typical impedance of 2-winding transformer can be found when %Z & X/R is selected; this is found by the American National Standard.

The transformer steps down to 33 kV using rated 100 kVA, this occurs when the grid is connected to the 132 kV. On the other side, the transformer can be found to be connected to Bus 2 where Load 1 is also connected. Connected to Bus 3, another transformer can now be found that is stepped down from 33 kV to 11 kV. Also with regards to this bus, a transformer 10 MVA rated is found in connection to Bus 4 (0.44 kV). Load 3 and capacitor are both connected to the 0.44 kV bus. The phase shift is fixed to 30° and all the connections of the network are star-delta.

5.2.3 The Harmonic Source

If the static load source contains significant current harmonic distortion, it can be modelled and regarded as a current harmonic. Also, users can specify voltage or current spectrum this is possible because of ETAP. Looking at Bus 3 and Load 2 (static load), it can be connected, and contains the harmonic source. When selecting Harmonic page of the load, it is then possible to specify the harmonic source. The harmonic source in the system is represented as a 12-pulse.

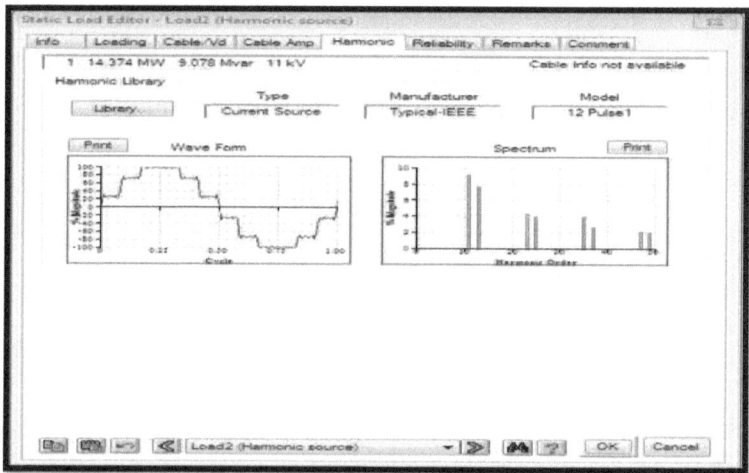

Figure 33 Load 2, Harmonic Page

The IHD is fixed to 3% and the default setting of harmonic limits made by ETAP, and the THD is fixed at 5%. The stated value will be used in comparison with the calculated THD from the Harmonic Load Flow calculation and all surpassing this limit will end in an alert appearing in the resulting report. The exceeded value of THD and IHD are not the only things that will appear on the report, since an over-voltage and under-voltage condition will also appear. The margin flag can be resolved by using load alternation, and the critical flag in the report means that it is crucial to cover the problem soon.

5.3 Results of Balanced Load Flow Analysis

The values obtained from the load flow are used in order to tune the harmonic filter, making these obtained values of great importance. (MVA) and existing system power factor are the two pieces of information that are needed. Additional parameters can be shown, for instance (MW+jMvar). For the current time, all displayed parameters are to be used later on in the calculation.

Power System Harmonic Analysis Using ETAP

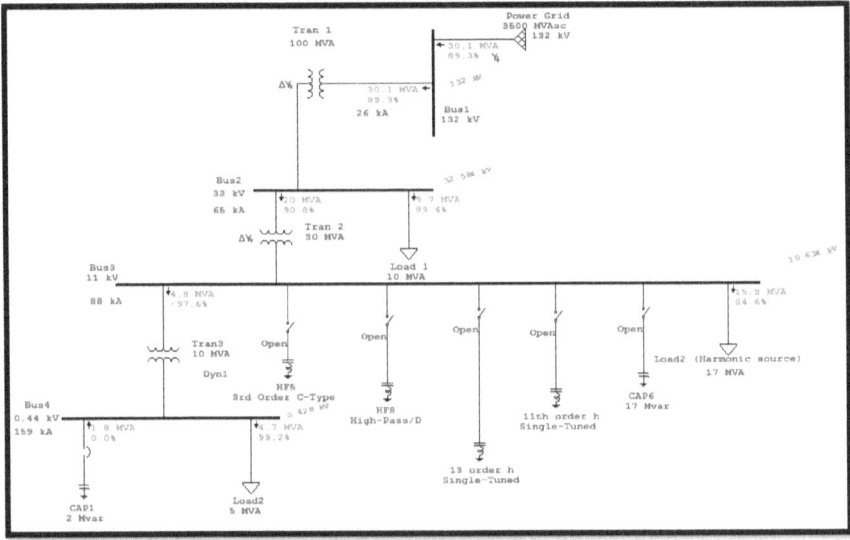

Figure 34 Balanced Load Flow Results

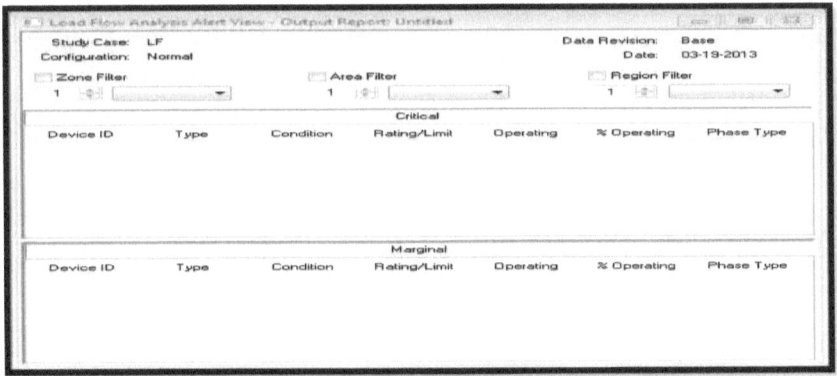

Figure 35 Load Flow Analysis Alert View

5.4 Results of Harmonic Load Flow Analysis

The results that have been obtained from harmonic analysis represent the THD in each bus, and also show the (rms) voltage. Notified in the report are any and all violations at any point in the network. The harmonic source is located at Bus 3 where it has the most distorted bus.

The THD should not exceed 5% and it is found to reach 11.60%. The Appendix in the end of this report contains further information on harmonic analysis.

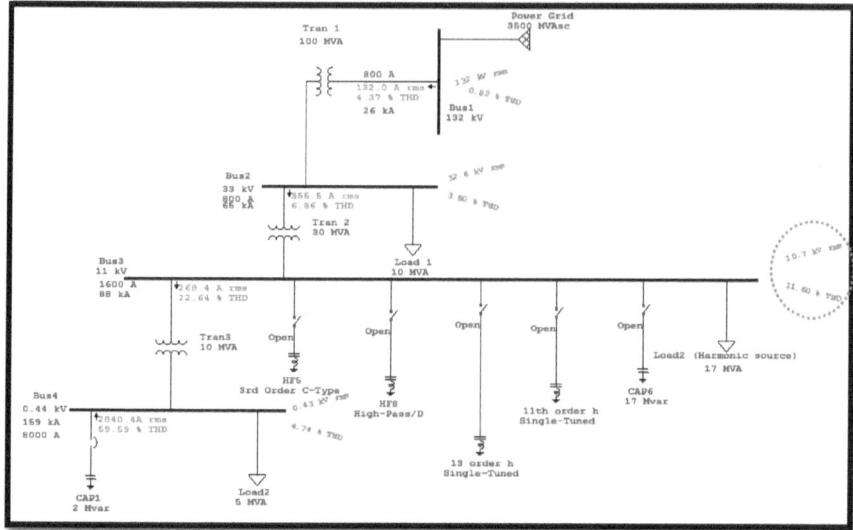

Figure 36 Harmonic Analysis Results

At The Harmonic Load Flow Analysis, all buses in the network are shown the IHD and THD by clicking on Alert View. It is important to note any and all alerts that are critical, so action must be taken in order to solve the problem at hand. In table 2, the critical harmonic that should be taken away from the system is illustrated.

Harmonic Order	Harmonic Location
11th	Bus 3 and Bus 4
13th	Bus 3
23th	Bus 3
25th	Bus 3
35th	Bus 3

Table 2 Harmonics order

Power System Harmonic Analysis Using ETAP

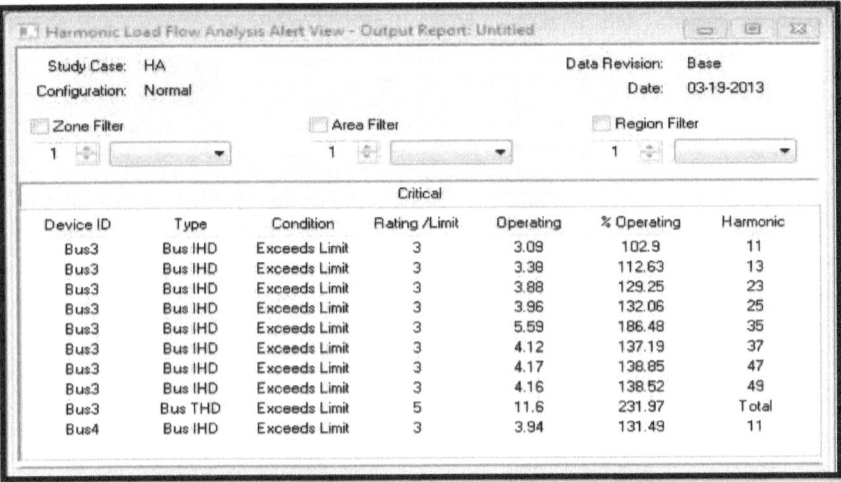

Figure 37 Harmonic Analysis Alert View

As seen in (Figure 37), harmonics are present in the system from the 11th to 49th order. These harmonics are generated by the harmonic source. The 11th and 13th order harmonics are the two main harmonics. They must be cleaned because they are near the fundamental frequency and resulting distorted current and voltage waveforms. The voltage waveforms and voltage spectrum are plotted for all buses. The most distorted bus is Bus 3 (11kV) due to the connected harmonic source.

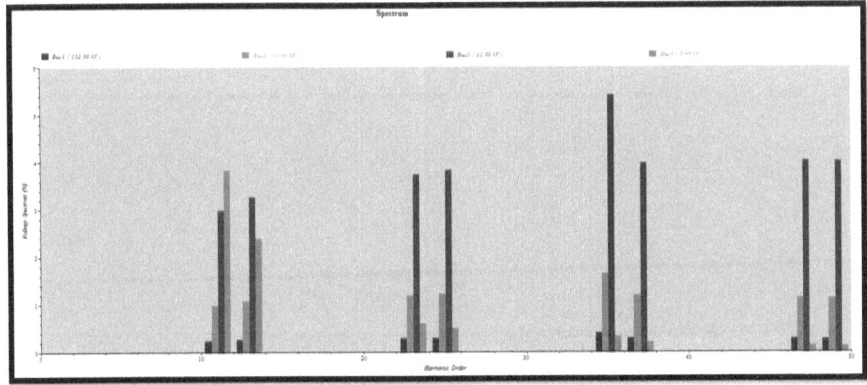

Figure 38 Harmonic Analysis Plot all buses (the voltage spuctrum)

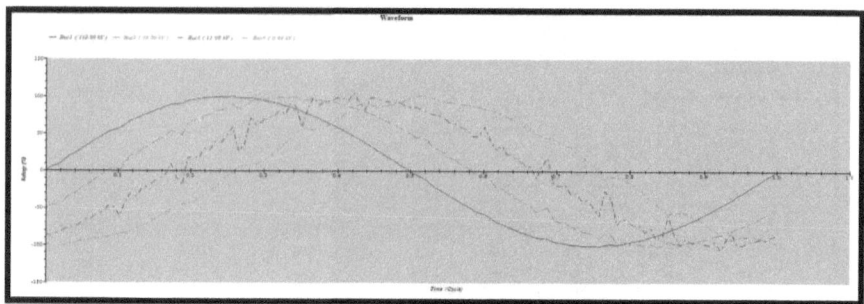

Figure 39 Harmonic Analysis Plot all buses (the voltage wafeform)

The graph below shows the voltage spectrum with the percentage of voltage spectrum against the Harmonic Order. At Bus 3, 11th, 13th, 23rd, 25th, 37th, 47th and 49th are the harmonic orders.

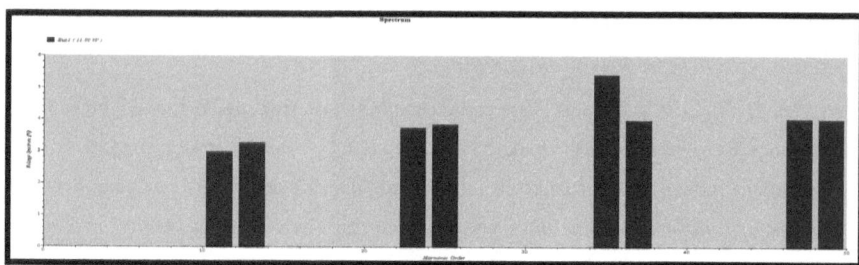

Figure 40 Bus 3 Spectrum before the elimination

As it can be seen below, the voltage waveform and voltage spectrum for Bus 3 are plotted to see the most distorted bus clearly.

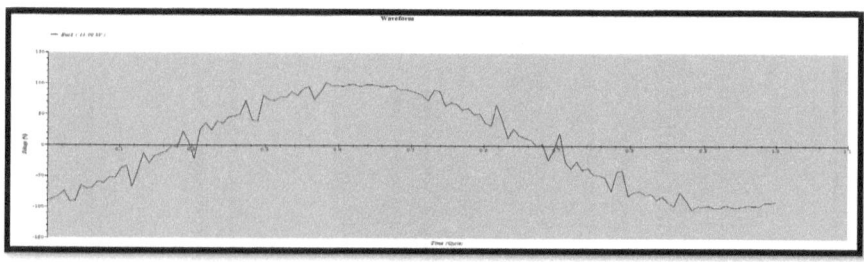

Figure 41 Harmonic Plot Bus 3

5.5 Results of Harmonic Scan

The impedances and phase angles can be found displayed throughout the network, from the frequency scan. Checking for if there are any resonance problems that might affect the network is why running this study is important. Because of the existing presence of the capacitance and inductance, there are also existing possibilities of resonance conditions regarding this network. The capacitor reactance and the reactance of the network will result in cancelling each other out. However, in some cases the resonance condition is not harmful to the system and this is the case when no critical alerts are found in the load flow analysis. Also, the resonance appearing in the frequency scan is between 6th and 7th order and the harmonic source at load 2 is 12-pulse rectifier so the lower harmonic order that can be generated is 11th. As a result, the harmonic source is not present at resonance frequency this distances the network from all resonance problem. If there are any parallel resonance conditions present, it is possible for a filter to be tuned at a particular harmonic order.

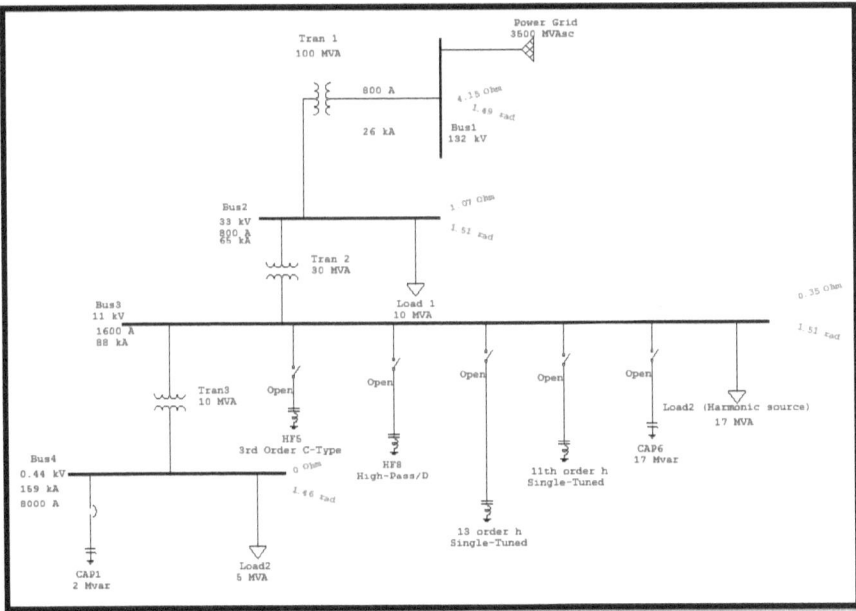

Figure 42 Result of Frequency Scan

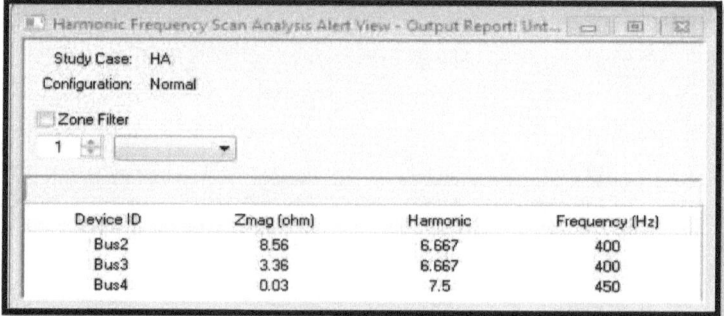

Figure 43 Harmonic Frequency Scan Alert View

The curves for the impedance and the phase angle need to be plotted for all buses and Bus 3 need to be plotted individually to carry out the frequency scan analysis.

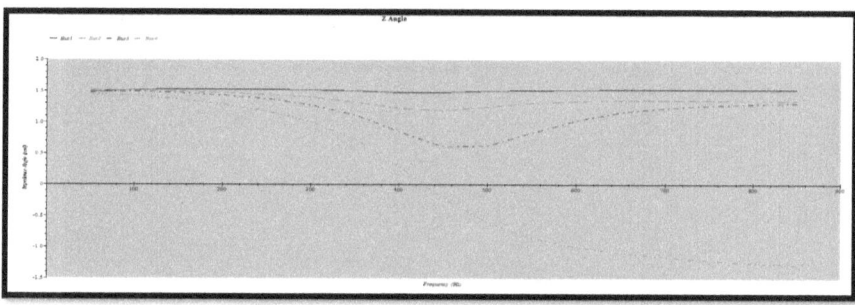

Figure 44 Harmonic frequency scan, impedance angles (all buses)

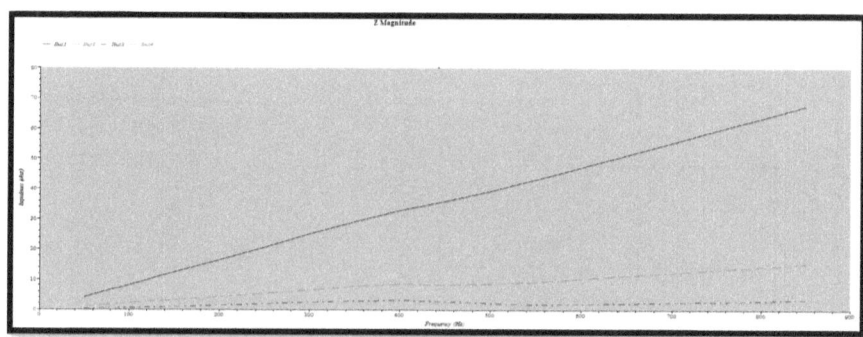

Figure 45 Harmonic frequency scan, impedance magnitude (All buses)

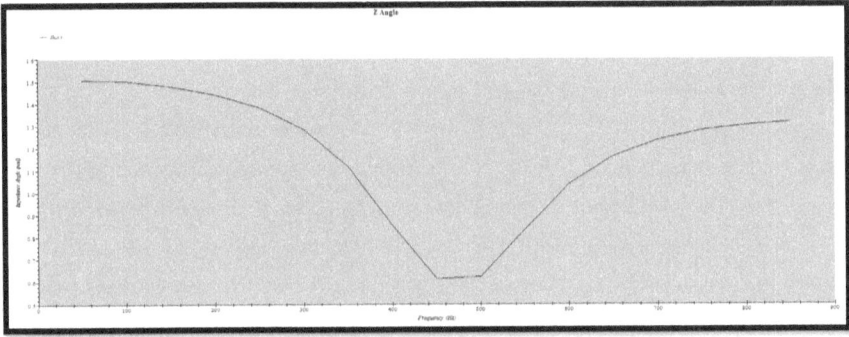

Figure 46 Harmonic frequency scan, impedance magnitude (All buses)

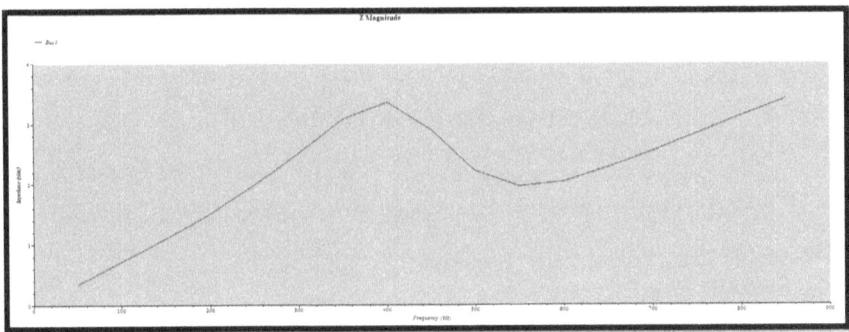

Figure 47 Harmonic frequency scan, impedance magnitude (Bus 3)

Harmonic load flow and harmonic frequency scan calculations are only possible after having modelled the components in the network and having found the results from the load flow. This is the time to benefit and use the results obtained in order to find a way to remove the harmonics and retain the THD to its limit. There are many filters that exist that can reduce harmonics; the challenge lies in finding which one is the best.

5.6 Harmonic Elimination

5.6.1 Harmonic Elimination Using Capacitor Bank (CB)

The Bus 3 in the network does not supply many feeders, in this case the capacitor bank will have the same effect as a low pass filter after connecting it at the low voltage side of the transformer (Tran 2). If Bus no.3 had many feeders a low pass filter would be connected instead, due to the fact that the low pass filter has a line reactor and voltage regulator connected at the beginning of the feeder to isolate the portion of the system that may face high voltage. In our network, a capacitor bank will be used. The size of the capacitor needs to be precise in order to give the desired cut-off frequency when combined with the system impedance and transformer leakage inductance. A 17 Mvar is connected at Bus 3 to provide reactive power compensation, power factor correction and reduce losses in the system. As a result, the THD will be reduced from 11.60% to 1.32% at Bus 3 and Bus 4 reduced to 1.47%. Before and after the capacitor bank distortion appears at the capacitor CAP 1 but this will not affect the system, only the upstream flow, which shows the clean action of the capacitor bank.

- Finding the optimal capacitor size

From load flow analysis (Figure 48) we can find the maximum and operating values of Apparent power (S) for transformer 2 (Tran 2), which is connected to Bus 3 where the harmonic source exists. The value of Active power (P) can also be obtained.

In order to find the value of Reactive power (Q) for the capacitor, we need to find the difference of Reactive power between two points. First, when Tran 2 operates at its maximum (30 MVA). Second, when Tran2 operating normally (20.016 MVA). The Active power is also needed which is 18.2 MW.

From $\quad S^2 = P^2 + Q^2$

$$Q = \sqrt{S^2 - P^2} \qquad \text{At first point } Q1 = \sqrt{(30M)^2 - (18.2M)^2} = 23.8 \; Mvar$$

$$\text{At second point } Q2 = \sqrt{(20.016M)^2 - (18.2M)^2} = 8.3 \; Mvar$$

The capacitor rating will be $Q2 - Q1 = 23.8 - 8.3 = 15.5 \; Mvar$ (which almost 17 Mvar).

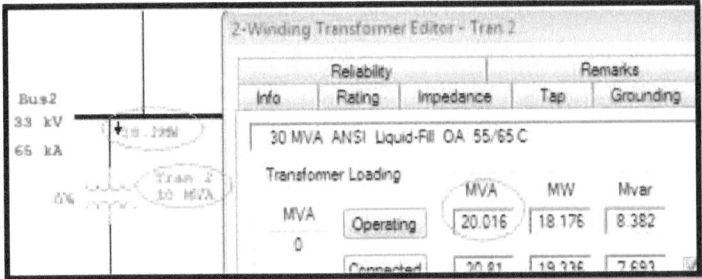

Figure 48 Load flow analysis on left and Transformer Edit page for Tran 2 on right

- The optimal capacitor placement

In order to obtain the maximum benefits from connecting the capacitor bank, it is suggested to install it as close to the inductive reactance load (Mvar) as possible. Practically speaking, the economic and availability of capacitor (Mvar) sizes is considered. As result, the capacitor can be located at different places near the load. An experiment in the field suggested locating the capacitor bank within a distance of 1/2 to 2/3 of total line length. Yet, using a program model such as ETAP can identify and place the capacitor at an optimal location considering the largest size that can be installed to minimize the voltage change capacitor switched. The computer program can find the optimal place by installing the smallest capacitor size that the system uses in each line of each feeder and then the total losses in the circuit can be determined. Therefore, the computer can select the best place that has the lowest losses [18].

The Optimal Capacitor Placement (OPP) feature in ETAP software at Brunel University was not purchased, so that in this project an estimation of location is used based on the information above.

Power System Harmonic Analysis Using ETAP

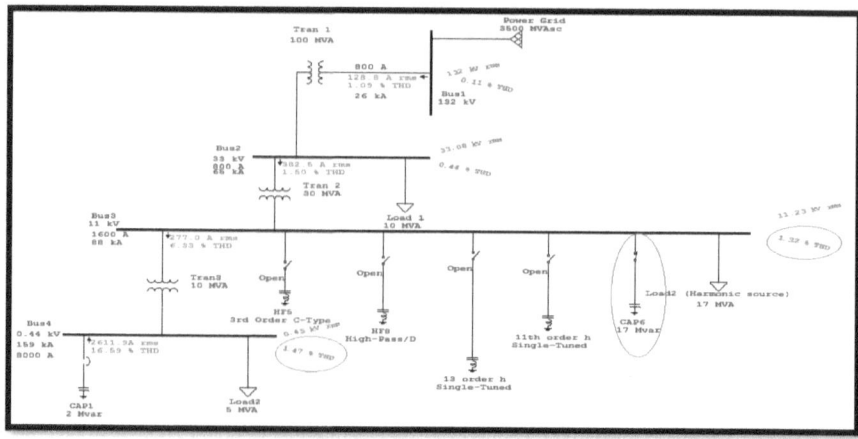

Figure 49 Harmonic Analysis after connecting capacitor bank

The following graphs show the difference before and after connecting the capacitor bank in terms of voltage waveform, spectrum, impedance (Z) magnitude and impedance (Z) angle. As a result, the capacitor bank has worked properly to reduce the THD in the system. Please refer to the Report Manager in the Appendix generated by ETAP for further analysis.

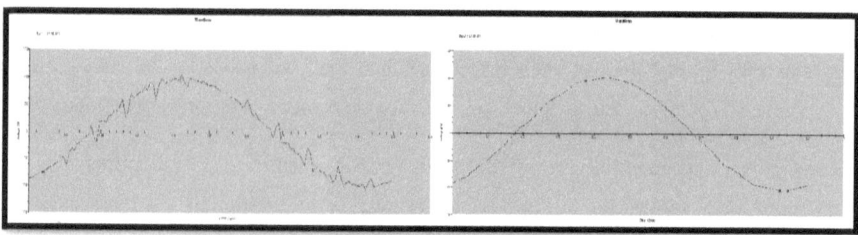

Figure 50 Voltage waveform before and after CB at Bus 3

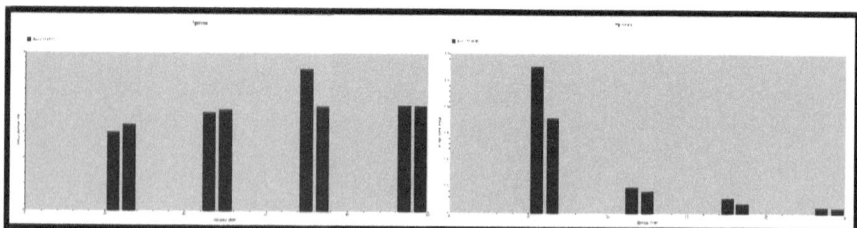

Figure 51 Spectrums before and after the CB at Bus 3

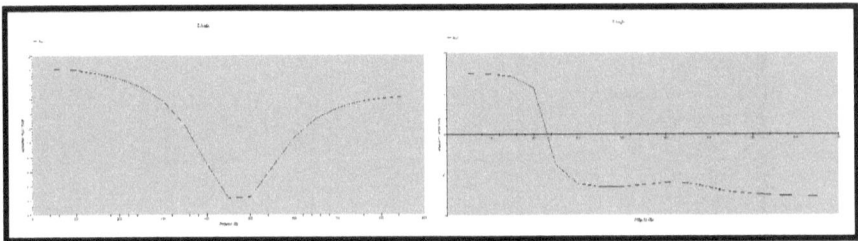

Figure 52 Frequency scan, Z angle before and after CB at Bus 3

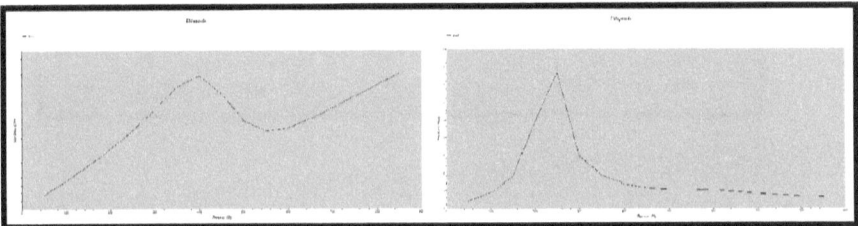

Figure 53 Frequency scan, Z magnitude before and after CB at Bus 3

5.6.2 Harmonic Elimination using C Filter

The C filter can be found in the ETAP library. This type of filter can perform the same job as a low pass filter to reduce multiple harmonics in the system. ETAP assumes the filter is installed in a single phase so the values of the KV and Kvar must be entered as a single phase.

In order to design the C filter we need to:

1. Obtain the Bus 3 voltage (KV) from Load Flow Analysis and divide it by $\sqrt{3}$, because the filter is connected in a wye-connection.

At Bus 3 KV=10.733K, single phase equal to $\frac{10.70K}{\sqrt{3}} = 6.2$ KV

2. Obtain the sum of Mvar at Bus 3 and then divide it by 3.

The values are 6.8M+7.6M = $\frac{14.4M}{3}$ = 4800 Kvar (about 5000Kvar)

3. After entering these values in Parameter page of harmonic, the value for the capacitor will be calculated automatically.

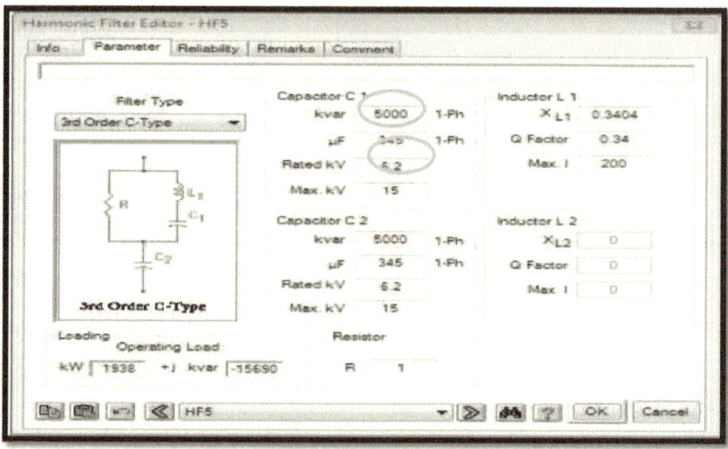

Figure 54 C Filter Parameter page

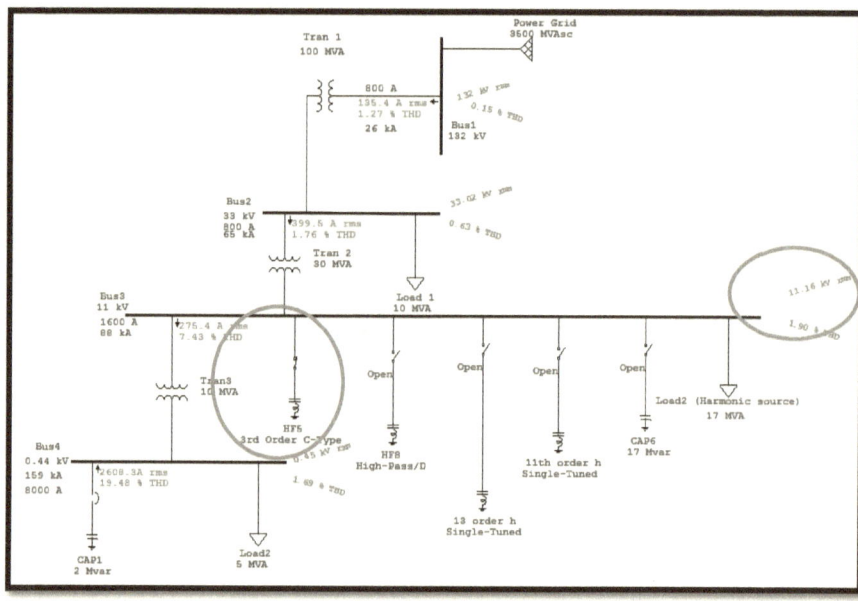

Figure 55 The Network after connecting C Filter

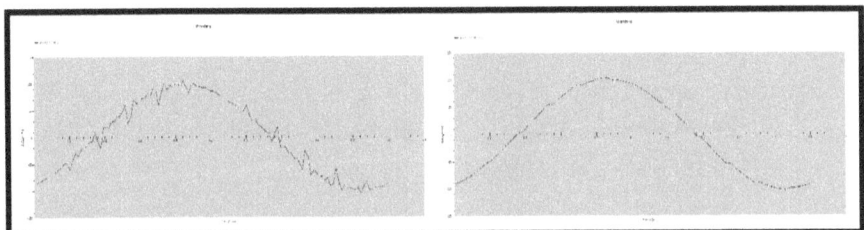

Figure 56 Voltage waveform before and after C filter

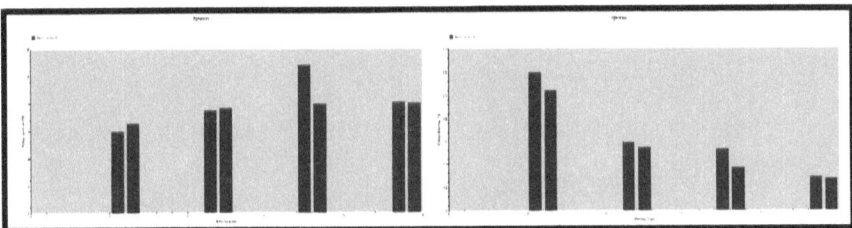

Figure 57 Spectrum before and after C filter

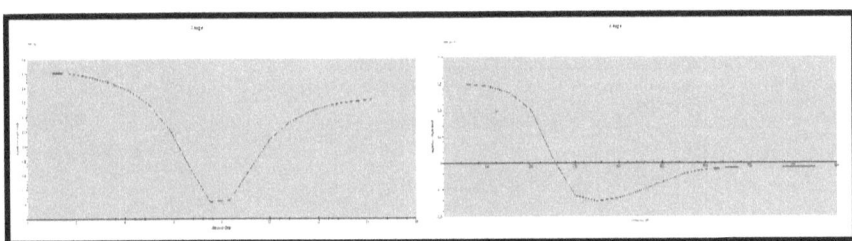

Figure 58 Frequency scan, Z angle

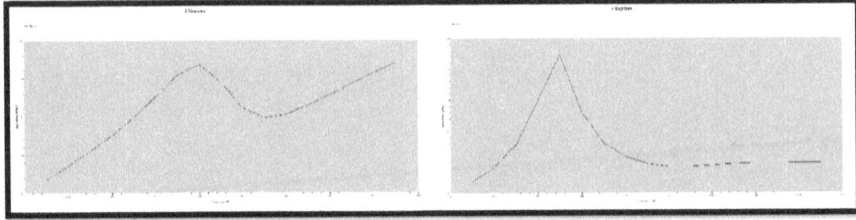

Figure 59 Frequency scan, Z magnitude

5.6.3 Harmonic Elimination using Single Tune Filter + High Pass filter

A Single-Tuned harmonic filter is selected from the ETAP library. This filter can provide low impedance for a specific harmonic current in order to minimize the impact generated by the non-linear load. This network has 2 major harmonics which are 11th, 13th. The other harmonics in the system can be eliminated using a high pass filter.

There are some issues that should be considered at the time of designing notch filter:

4. The filter should be tuned slightly lower than the harmonic order to be filtered.

5. Add the filter to the system starting with lowest significant harmonic exist in the system to avoid a resonance problem.

6. Consider the capacity of the bus at the design level to keep the filter away from any excessive duty.

In order to start designing filter for harmonic order 11th, the steps needed are:

1. Obtain the harmonic current of the 11th order from the Harmonic Order Slider. This can be done by running the Harmonic Flow Analysis. Then, select the 11th order from the Harmonic Order Slider. In this case is equal to 42 A.

2. Obtain the values of MVA and PF from the Load Flow Analysis, which 15.9 MVA and PF of 84.6%.

From the Parameter Page of Harmonic Filter, a size Filter button should be selected to put all the values obtained. Since this filter also works as a power factor correction, it is possible to choose the Desired PF which was chosen as 95%. The Q factor for the Inductor L1 can be calculated by

$Q = \frac{nXL}{R}$, where n = harmonic order, X_L = value of inductor in Ω, R= external resistance

$Q = \frac{11 \times 0.34}{1} = 3.74$

Power System Harmonic Analysis Using ETAP

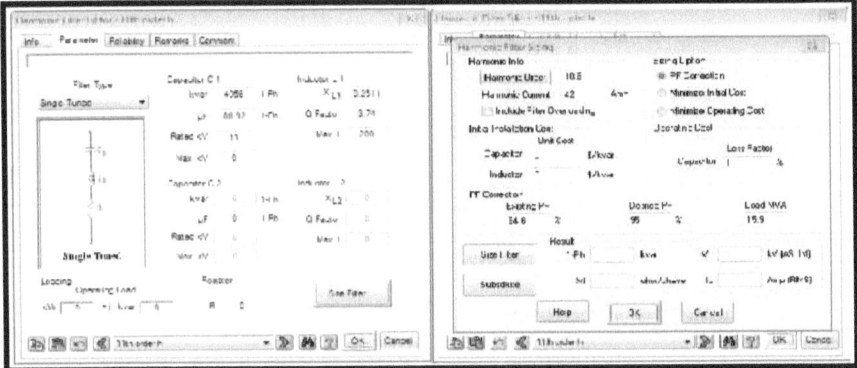

Figure 60 Harmonic Parameter and Filter Sizing Pages

The same procedure can be followed to design a filter for the 13th harmonic order. The following is a description of the results of installing these two single-tuned filters. Practically, all harmonics above the 15th order can be mitigated by using a low pass filter or any filter that can reduce widespread harmonics such as damped high pass filter. This is because the harmonics above the 15th order are far from the fundamental frequency. If a single tuned filter is used in this network for the 11th, 13th and 35th orders, the THD in the system will be reduced, but not as much as using a single tuned filter for harmonics up to the 15th order and a low pass filter for above this order.

Power System Harmonic Analysis Using ETAP

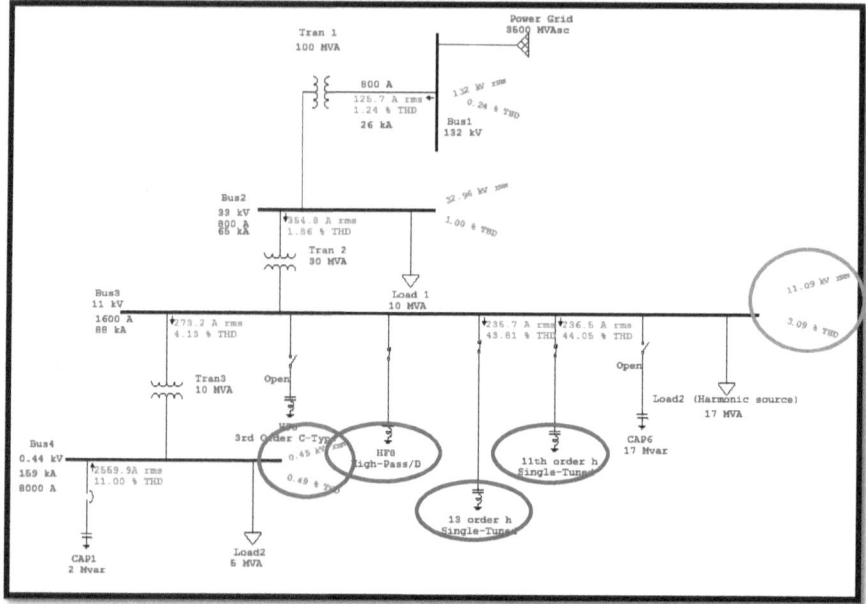

Figure 61 Harmonic Analysis after connecting Single Tune Filter

The THD at the distorted Bus 3 is now below the limit specified. This means that installing a filter for each harmonic order has worked successfully.

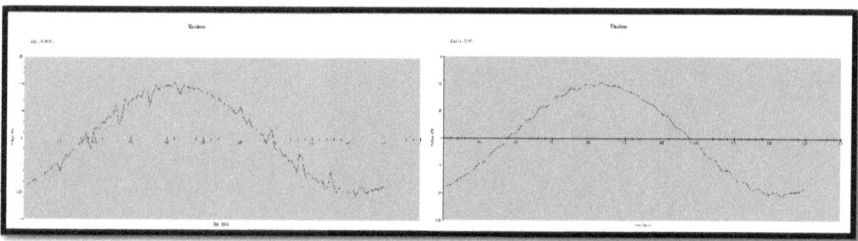

Figure 62 Voltage waveform before and after notch filter at Bus 3

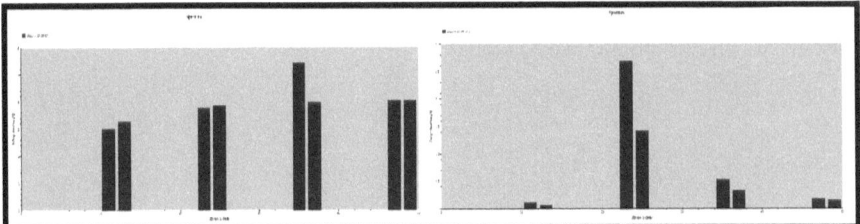

Figure 63 Spectrum before and after notch filter at Bus 3

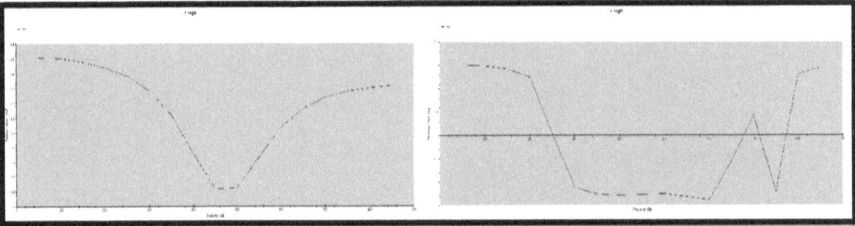

Figure 64 Frequency scan, Z angle before and after notch filter at Bus 3

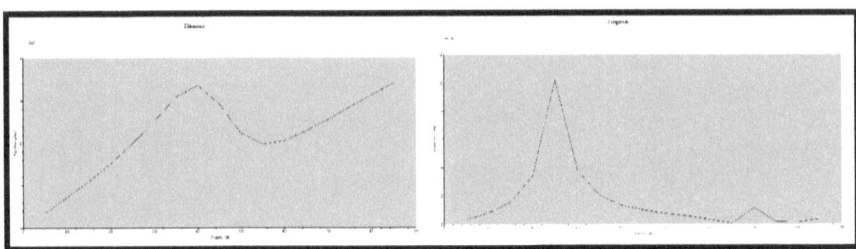

Figure 65 Frequency scan, Z magnitude before and after notch filter at Bus 3

5.7 Summary

The voltage total harmonic distortion was set at 5% THD, as in (Figure 37). The main concern in this chapter was to keep the distortion caused by the non-linear load from exceeding this limit. All three solutions were able to keep the distribution network away from any excess in THD. The most implemented approach is to install a compensate filter that consists of two or more brunches of single tuned filters tuned at low order harmonic (in our case the 11th and 13th) and a high pass filter tuned at higher frequencies. This configuration is used most in residential, domestic, and industrial non-linear loads [23]. The following table shows the reduction in THD at each solution.

Without Filters	With Capacitor Bank	With C Filter Type	With Single Tuned + HPF
11.60 % THD	1.32 % THD	1.90 % THD	3.09 % THD

Table 3 Reduction in THD in each solution

Chapter 6

Conclusion and Limitation

6.1 Conclusion

Electrical Transient Analysis Program (ETAP) is used to analyse harmonics in electrical power networks. The two main features of ETAP aimed at analysing harmonic in ETAP are Harmonic Load Flow (HLF) and Harmonic Frequency Scan (HFS). The HLF is used to determine the IHD and THD in each bus that caused by non-linear load, where HFS is used to check if the network contains any resonance conditions that may result in overvoltage. ETAP uses IEEE Standards 519 to model its components and specify the percentage of IHD and THD that should not be exceeded.

A power network of four buses is used to apply harmonic analysis studies. The harmonic source was a Typical-IEEE 12-pules rectifier. This rectifier is modelled as non-linear load at Bus no.3. After running the HLF and HFS, harmonics were found in the system as a result of this rectifier. Action must be taken to clean out the system from any harmonics. In this report three methods of harmonic mitigation are used which are a capacitor bank, notch filter and C filter, where each one of these methods has its advantages and drawbacks.

Connecting the capacitor bank at the proper place can improve the system's power factor, reduce the system losses and improve voltage profile, all these factors contribute to reduce the THD of the system, yet the most important issues concerning this method are the capacitor sizing and location. On the other hand, the capacitor bank sometimes causes resonance problems such as high voltage and current and possible magnification of capacitor switching. Overvoltage will be obtained if the switched capacitor is used.

Passive filters in general are also common used in industry to reduce harmonics due to their simplicity and low cost. The most economic passive type is the shunt passive filter (notch), used in this project to reduce harmonics. The benefits obtained from installing a notch filter are to provide a low impedance path for a particular harmonic current through the filter in order to mitigate this particular harmonic order, it also provides power factor correction. The problem with this type of filter is the number of the notch filter used in order to get the optimum result of harmonic reduction. It is very likely to design, for example, a filter for the 11th if the 13th order exists in the system. In other words each harmonic order requires a single filter. As result, this type of filter is not effective for multiple harmonic frequencies.

A C filter is a good option for the reduction of harmonics in industries or utility systems when considering multiple harmonic frequencies. This type of filter can ease a varying range of steady state and time varying harmonic and inter-harmonic frequencies that may be caused by any type of electronic converters.

Because an active filter is a new technology, its cost is higher than a passive filter. Yet, in the future active filters will become cheaper due to their size, fast response and ability to reduce widespread harmonics. At this time passive filters are the cheapest available in the market. Their problems include the huge size of capacitors and their dependence on system impedance.

Passive filters, then, are used in this power network to eliminate harmonics in the voltage and current waveform, improve power factor and reduce harmonic power losses. Generally, the problem with the passive filter is the size of the capacitor, which is huge, as well as the fact that the filter depends on the system impedance. However, it is very reliable and it can operate for a long time with no need for maintenance.

6.2 Limitation

The harmonic analysis carried out using ETAP had considered harmonic reduction alone, which is not cost effective in some cases. For example, using the capacitor bank for power factor correction is only focus on reducing THD. But in reality it is very costly. In order to obtain the optimum solution, the cost and solution methods must both be considered.

All filters used in ETAP software are passive filters. An active filter should be applied in the ETAP library because active filter implementation in some cases is much better than passive filter use, due to factors such as the good response of active filters to changing load and harmonic variation. In addition, it is very hard to design an active filter using existing components in ETAP.

REFERANCE:

[1] Harm_Intro, (2011), *An introduction to power system harmonics*. [On Line] Available at: http://www.powerstudies.com/articles/Harm_Intro.pdf [Accessed: 23 October 2012].

[2] Product Data Bulletin (1994), *Power System Harmonics Causes and Effects of Variable Frequency Drives Relative to the IEEE 519-1992 Standard*. [On Line]. Available at: http://www.alamedaelectric.com/Modicon%20Documents/AC%20Drive%20Power%20System%20Harmonics.pdf [Accessed: 21 November 2012].

[3] Dr. Mohamed Darwish (2012), *Power Electronics and FACTS* [Lecture presented to MSc Sustainable Power, Brunel University.

[4] Muhammad H. Rashid. *Power Electronics: Circuit Devices and Applications*, third edition. 2004 PP 23-395.

[5] ETAP (2012), Harmonic *Analysis Software*. [On line] Available at: http://www.etap.ca/products/harmonic-analysis-software/ [Accessed: 11 November 2012].

[6] Help (2012), *ETAP Tutorials - Training Videos & PDF Documents*. [On Line] Available at: http://etap.com/training/tutorials-training-videos.htm [Accessed: 14 November 2012]

[7] ETAP Software (2012), *Products Overview*. [On Line] Available at: http://etap.com/electrical-power-system-software/etap-products.htm [Accessed: 21 November 2012].

[8] ETAP help (2012), *ETAP 7.5.2 software*, Phoenix, AZ.

[9] Electronics-tutorial (2012), *Electronics Tutorial about Passive Low Pass Filters* [On Line] Available at: http://www.electronics-tutorials.ws/filter/filter_2.html [Accessed: 14 December 2012].

[10] Kerry Lacanette (2010), National Semiconductor Application Note, A Basic Introduction to Filters—Active, Passive, and Switched-Capacitor. [On Line] http://www.national.com/an/AN/AN-779.pdf [Accessed: 25 November 2012].

[11] Muhammad H. Rashid (2004), *Power Electronics: Circuit Devices and Applications*. Third edition.

[12] Ned Mohan, Tore M. Undeland, William P. Robbins, *Power Electronics: Converters, Applications, and Desi.* third edition. 2003.

[13] Daniel W. Hart (1997), *Introduction to Power Electronics.*

[14] Product Data Bulletin (1994), *Power System Harmonics Causes and Effects of Variable Frequency Drives Relative to the IEEE 519-1992 Standard.* [On Line] Available at: http://static.schneider-electric.us/docs/Motor%20Control/AC%20Drives/8803PD9402.pdf [Accessed: 14 October 2012].

[15] ETAP Software (2012*), Optimal Capacitor Placement Software.* [On Line] Available at: http://etap.com/distribution-systems/optimal-capacitor-placement.htm [Accessed: 11 December 2012].

[16] Ali M.Eltamaly (2009). *Harmonics Reduction Techniques in Renewable Energy Interfacing Converters.* [On line] available at http://www.intechopen.com/source/pdfs/9327/InTech-Harmonics_reduction_techniques_in_renewable_energy_interfacing_converters.pdf [accessed: 17 October 2012].

[17] Technocommerical (2005). *Harmonics-A power Quality Problem.* Available at http://www.mecoinst.com/media-releases/documents/harmonics.pdf [accessed: 14 Novmber 12].

[18] ETAP (2010). Harmonics analysis. [document presented by supervisor].

[19] ETAP (2011*). Harmonics Load Flow Software.* [on line] Available at http://etap.com/harmonic-analysis/harmonic-analysis.htm [accessed: 17 Novmber 2012].

[20] ETAP PowerStation (2009). *Harmonics analysis.* [on line] Available at http://www.powerserv-ei.com/web_documents/chapter_21_-_harmonic_analysis.pdf [accessed: 17 December 2012].

[21] Stephen David Hearn, PE (2010). *Basic Understanding of Harmonics in Electrical System.* [on line] available on

http://www.hearneng.com/WhitePapers/Understanding%20of%20Electrical%20Harmonics.pdf [accessed: 18 December 2012].

[22] J.L. Hernández, MA. Castro2, J. Carpio2 and A. Colmenar (2009). *Harmonics in Power System*. [on line] available at http://www.icrepq.com/ICREPQ'09/P1.pdf [accessed:19 January 2013].

[23] Fuchs E, Masoum M, (2008). *Power Quality in Power Systems and Electrical Machines*. Burlington: Elsevier press, pp 40.

[24] Frost & Sullivan (2003). *Harmonic Filters Overview*. [on line] available on] http://powersupplies.frost.com/prod/servlet/market-insight-top.pag?docid=SBRD-5LLMAB&ctxixpLink=FcmCtx7&ctxixpLabel=FcmCtx8 [accessed:22 January 2013].

[25] Dugan R, McGranaghan M, Santoso S, Beaty H (2002). 'Applied Harmonics'. *Electrical Power Systems Quality*. 2nd ed. McGraw-Hill, pp 252-264.

[26] L Morán, J Dixon, J Espinoza, R Wallace (No date). *Using Active Power Filters To Improve Power Quality*. [On line] available on http://web.ing.puc.cl/~power/paperspdf/dixon/37a.pdf [accessed : 23 February 2013].

[27] ETAP (2012). *ETAP Users*. [on line] available on http://etap.com/industries/etap-users-industry.htm [accessed: 9 February 2013].

[28] ETAP Power System Software (2012). *ETAP Product Literature Fact Sheets*. [On line] available on http://etap.com/demo-section/literature.htm [accessed: 12 February 2013].

[29] Chopade, P; Bikdash, M (2011) Minimizing Cost and Power Loss by Optimal Placement of Capacitor Using ETAP. *System Theory (SSST)*. **IEEE 43rd**, pp 24-29.

i want morebooks!

Buy your books fast and straightforward online - at one of world's fastest growing online book stores! Environmentally sound due to Print-on-Demand technologies.

Buy your books online at
www.get-morebooks.com

Kaufen Sie Ihre Bücher schnell und unkompliziert online – auf einer der am schnellsten wachsenden Buchhandelsplattformen weltweit! Dank Print-On-Demand umwelt- und ressourcenschonend produziert.

Bücher schneller online kaufen
www.morebooks.de

VDM Verlagsservicegesellschaft mbH
Heinrich-Böcking-Str. 6-8   Telefon: +49 681 3720 174   info@vdm-vsg.de
D - 66121 Saarbrücken   Telefax: +49 681 3720 1749   www.vdm-vsg.de

www.ingramcontent.com/pod-product-compliance
Lightning Source LLC
Chambersburg PA
CBHW031536210526
45464CB00003B/1029